AF820937

Acta Anatomica

International Journal of Anatomy, Embryology and Cell Biology

Founded 1945 by R. Chambers, G. Glimstedt, T. Peterfi and G. Wolf-Heidegger
Continued 1962–1974 by E.A. Boyden, 1955–1980 by A. Delmas,
1972–1980 by F. Walberg, 1945–1980 by G. Wolf-Heidegger,
1981–1988 by R. O'Rahilly, Davis, Calif., 1989–1990 by G.E. Goslow, Jr., Providence, R.I.,
1981–1992 by W. Lierse, Hamburg

KARGER

Printed in Switzerland
on acid-free paper by
Reinhardt Druck, Basel

Appears monthly:
3 volumes per year (12 issues)

Instructions to Authors
http://www.karger.ch/journals/aan/aaninst.htm

Acta Anatomica

Aims and Scope
Acta Anatomica is an international forum for communication among scientists interested in morphology at all levels of organization, from subcellular to macroscopy, and in morphogenesis, with emphasis on humans and higher vertebrates. The journal offers rapid exchange of novel and significant information gained from experimental and theoretical investigations on all aspects that are pertinent to the formation and transformation of biological structure.

Topics of special interest include: Vertebrate embryology (including developmental and reproductive biology), cell biology, neuroanatomy, functional and clinical anatomy (including application of modern techniques of imaging). Papers highlighting the links between structural and functional aspects are preferred.

Types of Papers
The journal publishes: (1) Original papers on experimental research. (2) Theoretical contributions. These will appear in the section 'Hypotheses and Concepts' and may include provocative hypotheses, provided that they are innovative, well founded and show promise for new avenues of research. (3) Short reviews/perspectives. While (1) and (2) are submitted contributions, reviews/perspectives (3) are invited. Suggestions concerning reviews on current topics of high interest are welcome.
All articles submitted must be original. All manuscripts must be approved by all authors, and this has to be certified by the first author with his signature. Manuscripts may not have been published previously either in whole or in part, except in abstract form, and may not be under consideration by any other journal. Reported studies using animal and human subjects must have been approved by the appropriate institutional review boards.

Submission
Manuscripts from North and South America should be sent to:

Prof. A.W. English, PhD
Emory University School of Medicine
Dept. of Anatomy and Cell Biology
Atlanta, GA 30322 (USA)
Telefax (1) 404 727–3677
E Mail: acta@anatomy.emory.edu

Manuscripts from Japan and South-East Asia to:

Prof. Dr. Osamu Ohtani
Department of Anatomy II
Toyama Medical and Pharmaceutical University
2630 Sugitani
Toyama, 930-01 (Japan)
Telefax (81) 764 34 5010

Manuscripts from all other countries to:

Prof. Dr. med. Dr. rer. nat. H.-W. Denker
c/o S. Karger AG
Editorial Dept. 'Acta Anatomica'
P.O. Box
CH–4009 Basel (Switzerland)
Telefax (41) 61 306 14 34

All manuscripts are subject to peer review. Submission of an article for publication implies the transfer of the copyright from the author to the publisher upon acceptance. Accepted papers become the permanent property of Acta Anatomica and may not be reproduced by any means, in whole or in part, without the written consent of the publisher. It is the author's responsibility to obtain permission to reproduce illustrations, tables, etc. from other publications.

Format and Arrangement
Manuscripts should be written in English and be as concise as possible. They should be submitted in quadruplicate (with four sets of original illustrations), typewritten double-spaced on one side of the paper, with a wide margin. Lines should be numbered on the margin.
Title page: Title (main title underlined), the authors' names, and the institute where the work was conducted, a short title for use as running head, the full address of the author to whom correspondence should be sent. Please also supply Phone, Fax, and E-mail numbers. If different, indicate in addition a full address for revisions and proof reading separately.
Key words: For indexing purposes provide a list of up to 5 key words.
Abstract: Each paper needs an abstract of up to about 20 lines summarizing the essential new informations communicated.
Subheadings: Introduction; Materials and Methods; Results; Discussion.
Small type: Paragraphs which can or must be set in smaller type (methods, etc.) should be indicated with a 'p' (petit) in the margin on the left-hand side.
Footnotes: Abbreviations used in the paper will be printed as a footnote on the first page. Please list them on a separate page of the manuscript. Otherwise, footnotes should be avoided. When essential, they are numbered consecutively and typed at the foot of the appropriate page.
Nomenclature: The nomenclature used should conform, wherever possible, to the current edition of the *Nomina Anatomica, Nomina Anatomica Veterinaria,* or to the *Nomina Anatomica Avium.*
Tables and illustrations: Tables and illustrations (both numbered in Arabic numerals) should be prepared on separate sheets. Tables require a heading and figures a legend, also prepared on a separate sheet. For the reproduction of illustrations, only good drawings and original photographs can be accepted; negatives or photocopies cannot be used. Due to technical reasons, figures with a screen background should not be submitted. When possible, group several illustrations on one block for reproduction (max. size 181 × 223 mm) or provide crop marks. On the back of each illustration, indicate its number, the author's name, and 'top' with a soft pencil.

Color illustrations: During the review process, the decision will be made as to which color illustrations of a manuscript can be reproduced free of charge. Authors will be informed of this at the time of acceptance of their paper. Otherwise, these illustrations will be reproduced at the author's expense. Up to 6 color illustrations per page can be integrated into the text at the reduced price of CHF 600.– / USD 545.00 per page.

References: References in the text should be made by *author's name* followed by the *year of publication,* arranged chronologically, then alphabetically. When there are more than two authors, use the first *author's name and et al.* When references are made to more than one paper by the same author, published in the same year, designate them as a, b, c, etc. The list of references should include only those publications which are cited in the text. Material submitted for publication but not yet accepted may be cited in the text as 'unpublished data' but must not be included in the reference list. In the list of references, arrange references *alphabetically* listing all authors, then year of publication. Abbreviate journal names according to Index Medicus.
Examples
(a) Journal papers: Kauffman, H.F., S. van der Heide, F. Beaumont, H. Blok, K. de Vries (1986) Class-specific antibody determination against *Aspergillus fumigatus* by means of the enzyme-linked immunosorbent assay. III. Comparative study: IgG, IgA, IgM ELISA titers, precipitating antibodies and IgE binding fractionation of the antigen. Int Arch Allergy Appl Immunol *80:* 300–306.
Hadler, W.A., S.R. Silveira (1981) Técnica histoquímica que permite discriminar o colesterol de substâncias afins e de seus ésteres. Cienc Cult 33(suppl): 586–587.
(b) Books/Monographs: Matthews, D.E., V.T. Farewell (1985) Using and Understanding Medical Statistics. Basel, Karger.
(c) Book chapters: Hardy, W.D. Jr., M. Essex (1986) FeLV-induced feline acquired immune deficiency syndrome: A model for human AIDS; in Klein, E. (ed): Acquired Immunodeficiency Syndrome. Prog Allergy. Basel, Karger, vol 37, pp 353–376.
Further details: Authors may consult our leaflet 'Rules for the Preparation of Manuscripts' available on request from the Publisher, S. Karger AG, P.O. Box, CH–4009 Basel (Switzerland), and the 'Guide to Writing Articles in English' by R. O'Rahilly (Acta anat. *131,* 1–2, 1988).

Cover illustrations: The Journal Acta Anatomica will publish cover illustrations relevant to any area of modern anatomical research. Cover illustrations will be chosen primarily on the basis of aesthetic appeal and not necessarily from articles in that issue of the journal. Authors or readers are encouraged to submit illustrations suitable for the cover. Only black-and-white illustrations can be accepted. Please submit two copies (max. 13 × 16 cm) to the appropriate editors. Submissions should include a brief caption on the back of the illustration.

Galley proofs: Unless indicated otherwise, galley proofs are sent to the first-named author and should be returned with the least possible delay. Alterations made in proofs, other than the correction of printer's errors, are charged to the author.

Reprints: Order forms and a price list are sent with the proofs. Orders submitted after the issue is printed are subject to considerably higher prices. Allow five weeks from date of publication for delivery of reprints.

KARGER
E-Mail karger@karger.ch
Fax + 41 61 306 12 34
http://www.karger.ch

General Information
http://www.karger.ch/journals/aan/aandes.htm

Acta Anatomica

Publication data: 'Acta Anatomica' is published 12 times annually. Volumes 155–157 with 4 issues each appear in 1996.

Subscription rates: Subscriptions run for a full calendar year. Prices are given per volume.

Personal subscription:

Subscription	+ *postage and handling*
CHF 420.60	+ CHF 14.70 Europe
CHF 420.60	+ CHF 22.40 Overseas
USD 323.40	+ USD 14.70

(Must be in the name of, billed to, and paid by an individual. Order must be marked 'personal subscription'.)

Institutional subscription:

Subscription	+ *postage and handling*
CHF 701.–	+ CHF 18.40 Europe
CHF 701.–	+ CHF 28.– Overseas
USD 539.00	+ USD 18.40

Airmail postage: CHF 24.–, USD 20.00. (Extra per volume.)

Microforms: Please direct all enquiries and orders to:
University Microfilms Inc.
300 North Zeeb Road, P.O. Box 1346
Ann Arbor, MI 48106-1346 (USA)

Subscription orders: Orders can be placed at agencies, bookstores, directly with the Publisher S. Karger AG, P.O. Box, CH – 4009 Basel (Switzerland) or with any of the following distributors:
Deutschland: S. Karger GmbH, Postfach
D – 79095 Freiburg
France: Librairie Luginbühl
36, bd de Latour-Maubourg, F–75007 Paris
Great Britain, Ireland: S. Karger Publishers
11 Ravenscroft Road
Beckenham, Kent BR3 4TP
USA: S. Karger Publishers, Inc.
26 West Avon Road, P.O. Box 529
Farmington, CT 06085
India, Bangladesh, Sri Lanka:
S. Karger India Subscription Agency
B – 5/132 Safdarjung Enclave
New Delhi-110029
Japan: Katakura Libri, Inc.
Take 1 Bldg, 2nd Floor, 4-5-7 Yushima
Bunkyo-ku, Tokyo 113

Change of address: Both old and new address should be sent to the subscription source.

Single issues and back volumes: Information can be obtained through the Publisher.

Advertising: Correspondence and rate requests should be addressed to the Publisher.

Bibliographic indices: This journal is regularly listed in bibliographic services, including *Current Contents®*.

KARGER
E-Mail karger@karger.ch
Fax + 41 61 306 12 34
http://www.karger.ch

Epithelial-Mesenchymal Transitions

Part II

Guest Editor
D. Newgreen, Melbourne

25 figures, 4 color plates, 2 tables, 1997

KARGER

S. Karger
Medical and Scientific Publishers
Basel · Freiburg · Paris · London
New York · New Delhi · Bangkok
Singapore · Tokyo · Sydney

Drug Dosage
The authors and the publisher have exerted every effort to ensure that drug selection and dosage set forth in this text are in accord with current recommendations and practice at the time of publication. However, in view of ongoing research, changes in government regulations, and the constant flow of information relating to drug therapy and drug reactions, the reader is urged to check the package insert for each drug for any change in indications and dosage and for added warnings and precautions. This is particularly important when the recommended agent is a new and/or infrequently employed drug.

Printed in Switzerland on acid-free paper by
Thür AG Offsetdruck, Pratteln
ISBN 3–8055–6435–X

Acta Anatomica

Vol. 156, No. 3, 1996

Contents

Part II

KARGER

E-Mail karger@karger.ch
Fax + 41 61 306 12 34
http://www.karger.ch

The list of contents is available at:
http://www.karger.ch/journals/aan/aancont.htm

Acta Anatomica

Contents

Part I

(Vol. 154, No. 1, 1995, previously released)

KARGER

E-Mail karger@karger.ch
Fax + 41 61 306 12 34
http://www.karger.ch

The list of contents is available at:
http://www.karger.ch/journals/aan/aancont.htm

Editor's Note

To believe in technology rather than in major concepts and theories is a trend prevailing right now in science, and thinking is often equalized with speculation. However, real progress in understanding nature does rely, I am convinced, on inspiration and on concepts that can form a basis on which reasonable experiments are to be planned. Clearly, the answers that we obtain depend on the questions we ask.

Epithelial-mesenchymal transition or transformation (EMT; both expressions are in use) is a concept that is attracting increasing interest. Like all concepts, it necessarily simplifies, and it provokes contradiction. On the other hand, EMT has turned out to be a theme that may be common to surprisingly many biological phenomena. The idea that EMT may be one of the fundamental processes that drive development has initiated the rethinking of many well-known processes in embryology. By now we see that the number of laboratories which are applying this concept is growing since it is found to be very useful in many more systems than was originally thought. Indeed, it is helping to open to our eyes to detect patterns of changes in molecules involved in morphogenesis. The postulate, prophetically formulated by Elizabeth Hay years ago, that certain regulatory genes may govern EMT similarly in various different systems has now attracted increasing interest. The idea could potentially be of considerable value for the investigator since it forces us to look over the fence of our favorite experimental system and of our preoccupations.

I am happy that Don Newgreen has so enthusiastically taken up my proposal to edit these two special issues of *Acta Anatomica* and to bring together contributions by major experts in this field. Issue I [Acta Anat 154/1/95] contained an introductory chapter on EMT by Elizabeth Hay, papers on involved molecules and general signalling events, and contributions on two of the major examples of EMT in embryology (neural crest and mesoderm formation). The issue at hand (Part II) contains contributions on involved molecular mechanisms and on several other biological systems in which EMT or the reverse processes, MET, play a role, including heart and kidney development, placentation and cancer progression. I would like to thank all contributors, and I invite anyone interested in the concept of EMT to use *Acta Anatomica* also in the future as a forum in which we will continue to put emphasis on this fascinating concept.

H.-W. Denker

See also 'Epithelial-Mesenchymal Transitions, Part I' (Acta Anat 154/1/95)

Donald F. Newgreen, Melbourne

Preface

Epithelium and mesenchyme constitute the two basic cell phenotypes, and these two states starkly contrast both in the extracellular relationships of the cell to adjacent cells and noncellular structures, and in the intracellular organisation of the cytoskeleton and metabolic apparatus. Yet, despite these fundamental differences, these two contrasting cell states are interconvertible.

By definition, the cells of metazoans must cling together: Evolutionarily, early forms may have consisted of epithelial-like structures with specialised, but still epithelial, infoldings. Although this meant that considerable control of the internal milieu was possible, to achieve the vast complexity of form and great range of sizes seen in the animal kingdom required cellular constituents between the epithelial cell layers. An expedient, early adopted in evolution, and used repeatedly in development, is simply the controlled conversion of certain epithelial cells into cells capable of occupying the subepithelial spaces, into cells capable of giving this space new structure and function, that is, into mesenchyme cells.

In the previous volume [Acta Anat 154/1/95], an overview of the field of epithelial-mesenchymal transitions was given by Hay. As a basic reference section for the biological systems described in detail both in the first volume and in the present volume, the first volume also presented reviews on the mechanisms at work in epithelial-mesenchymal transitions. The molecular mechanisms whereby cells interact adhesively with their cellular neighbors were described by Prieto and Crossin. The mechanisms whereby cells adhere to their noncellular surroundings were supplied by Garratt and Humphries. The complex interactive network of structural and signalling linkages between these cell-surface adhesive molecules and the intercellular structural and functional organisation of the cytoskeleton were elegantly presented by Geiger, Yehuda-Levenberg and Bershadsky.

Two archetypal embryonic epithelial-mesenchymal transitions, gastrulation in mammals and neural crest morphogenesis, were then reviewed by Viebahn and by Duband, Monier, Delannet and Newgreen, respectively.

In the present volume new ideas on the molecular control of epithelial-mesenchymal transitions via kringle domain proteases are given by Théry and Stern. Details of another epithelial-mesenchymal transition in embryogenesis are supplied by Markwald, C. Eisenberg, L. Eisenberg, Trusk and Sugi in heart morphogenesis. Conversions in the reverse direction, mesenchyme to epithelium, are also known: Davies describes such events in kidney development. Recent studies of implantation of the mammalian blastocyst have allowed this to be reinterpreted as an event similar to classical epithelial-mesenchymal conversion; this is described by Vićovac and Aplin.

The creation of a new cell state, mesenchyme, allowed an enormous step in the order and complexity of the metazoan body plan to be realised. However, this creation is a two-edged sword, for if such cellular conversions occur in an unregulated way and are coupled to abnormalities of cell growth, metastasis occurs in previously benign tumors. This most feared of disorders is explored by Birchmeier, Birchmeier and Brand-Saberi.

Finally, the molecular nature of epithelial-mesenchymal transitions can be examined in highly accessible and controlled tissue culture systems. This potentially allows for a greater range and depth of experimental manipulations, which may reveal fundamentals of the process obscured by system-specific details in natural examples. These model systems are described by Boyer, Vallés and Thiery.

The editors sincerely thank all the contributors for these valuable reviews.

Acta Anatomica

Acta Anat 1996;156:162–172

C. Théry
C.D. Stern

Department of Genetics and Development, College of Physicians and Surgeons of Columbia University, New York, N.Y., USA

Roles of Kringle Domain-Containing Serine Proteases in Epithelial-Mesenchymal Transitions during Embryonic Development

Key Words
Epithelial-mesenchymal transitions
Proteases
Kringle domains
Plasminogen
Plasminogen activators
Hepatocyte growth factor/scatter factor
HGF-like/macrophage-stimulating protein
Tyrosine kinase receptors

Abstract
Transformation of an epithelial sheet into a migrating mesenchymal cell population implies the destruction of the basal lamina underlying the epithelium, and the subsequent localized digestion of the extracellular matrix by the migrating cells. Proteases are involved in these processes. Among them, molecules containing both a serine protease domain and at least one kringle domain have been identified as possible important effectors. Interestingly, related proteins containing an inactive serine protease domain also seem to play a role, suggesting that the function of these molecules in epithelial-mesenchymal transformation is not confined to proteolytic digestion of cell attachments. Instead, these molecules act through specific tyrosine kinase receptors in the membrane of the responding cells. In this review, we summarize data implicating this family of molecules in various epithelial-mesenchymal transitions during embryonic development. Our major focus of attention are: hepatocyte growth factor/scatter factor (HGF/SF), its tyrosine kinase receptor proto-oncogene c-met, and the related peptide factor HGF-like/macrophage-stimulating protein (HGFl/MSP), whose receptor is the Ron tyrosine kinase. c-met and Ron also have another close homolog in the chick, called Sea, whose ligand remains unknown. Interestingly, HGF/SF is activated by other plasminogen-related molecules which, apart from a specific activator, include the protease urokinase. HGF/SF, c-met and HGF/MSP are expressed in dynamic ways during early embryonic development, correlating with regions undergoing epithelial/mesenchymal transformations. Moreover, several assays are now starting to reveal great pleiotropism of function during development, including both the loss and the acquisition of epithelial morphology according to the cell type and assay used, as well as angiogenesis, kidney tubule morphogenesis, cell motility, the maintenance of competence for neural induction and some aspects of the later development of the musculoskeletal and nervous systems.

Abbreviations used in this paper:
EMT = Epithelial-mesenchymal transition; HGF/SF = hepatocyte growth factor/scatter factor; HGF1/MSP = hepatocyte growth factor-like/macrophage-stimulating protein; Shh/Vhh = sonic hedgehog/vertebrate hedgehog; TGFβ = transforming growth factor; tPA = tissue-type plasminogen activator; uPA = urokinase-type plasminogen activator.

KARGER

E-Mail karger@karger.ch
Fax + 41 61 306 12 34
http://www.karger.ch

0001–5180/96/1563–0162$12.00/0

Prof. Claudio D. Stern
Department of Genetics and Development
College of Physicians and Surgeons of
Columbia University
701 West 168th Street
New York, NY 10032 (USA)

Transformation of an epithelial sheet into a migrating mesenchymal cell population implies destruction of the basal lamina underlying the epithelium, and subsequent localized digestion of the extracellular matrix by the migrating cells [Hay, 1995]. Proteases are involved in these processes. Among them, molecules containing both a serine protease domain and at least one kringle domain have been identified as possible important effectors. Interestingly, related proteins containing an inactive serine protease domain also seem to play a role. Here we summarize data implicating this family of molecules in various epithelial-mesenchymal transitions (EMT) during embryonic development.

The Family of Kringle Domain-Containing Serine Proteases

Structure

The 'kringle domain' is an 80 amino acid three-disulphide stabilized tertiary structure. The presence of at least one of these structures together with a serine protease domain defines a new subfamily of proteins. The family presently contains nine members, whose properties are summarized in table 1. Figure 1 shows a schematic model of the structure of one of these members as an example.

Most of these proteins are active serine proteases. Some of them, however, either have not been shown to have a real protease activity (apolipoprotein-a, McLean et al. [1987]), or have an inactive protease domain, due to the substitution of some of the amino acids of the catalytic triad (HGF/SF, Nakamura et al. [1989]; HGFl/MSP, Degen et al. [1991], Han et al. [1991]).

The activity of these proteins is tightly regulated: they are secreted as inactive zymogen, present in circulating blood or various tissues, which need to be activated by proteolytic cleavage at the so-called 'activation domain'. This domain is located between the N-terminal part of the protein, which contains the kringle and other domains, and the C-terminal serine-protease domain. Some of these proteins then lose most of the N-terminal domain (thrombin, β-factor XIIa, HGF activator), but in most cases, the two parts remain linked by a disulphide bridge: the active form is therefore a heterodimer. An unusual feature is that, in several cases, the protease responsible for this activation is another member of the family (e.g. uPA and tPA for plasminogen, or HGF activator for HGF/SF), which can also be activated by its activated substrate, leading to complex cascades of regulation. As an example, the cycle leading to plasminogen activation is shown schematically in figure 2.

Functions

Several of the kringle-containing serine proteases have been studied for their role in blood coagulation: the active form of prothrombin, thrombin, cleaves the soluble circulating fibrinogen into insoluble fibrin, leading to blood clotting. It also activates, by proteolysis, a seven transmembrane domain molecule present at the surface of blood platelets [Vu et al., 1991], the thrombin receptor, leading to the activation and subsequent aggregation of the platelets. Factor XII is one of the components of the cascade leading to prothrombin activation.

The active form of plasminogen, plasmin, is the effector of the counter mechanism: fibrinolysis, i.e. degradation of fibrin into soluble products, which leads to the solubilization of the blood clot. But its substrates are also more diverse, and it can degrade various extracellular matrix proteins [for review, see Vassalli et al., 1991]. Transformation of plasminogen into plasmin is achieved by the plasminogen activators: urokinase- (uPA) or tissue-type (tPA). uPA itself also seems able to cleave extracellular matrix proteins such as fibronectin [McGuire and Seeds, 1990; Quigley et al., 1987]. These properties are potentially important for EMT, and will be discussed in more detail below.

The functions of other members of the family are still poorly understood. Apolipoprotein(a), for example, is present in blood as a complex with lipids called lipoprotein (a), but no protease activation and no serine protease activity have been shown [McLean et al., 1987]. One of the most recently discovered members of the family, HGF activator, is until now known only for its ability to cleave the single chain precursor of HGF/SF into the active dimeric form [Miyazawa et al., 1993].

Finally, HGF/SF and HGFl/MSP, even if they both lack protease activity, have several biological effects on cell lines in vitro, notably a motogenic activity on epithelial cells (HGF/SF, Stoker et al. [1987]) or monocytes (HGFl/MSP, Skeel et al. [1991]), which make them interesting in the context of EMT. They will be discussed in more detail later.

Receptors

In addition to their substrates, some of the kringle-containing serine proteases have been shown to interact with specific cell-surface proteins, which act as their receptors. The N-terminal kringle-containing part is involved in this interaction. As expected from the variability of this part of the protein in the family, the receptors so far characterized are very different. The thrombin receptor is a ligand and substrate for thrombin, which lacks the kringle domains present in prothrombin. Thrombin receptor is therefore not

Table 1. Structure and properties of the nine members of the kringle domain-containing serine-protease family

Factor	Kringle domains	Other domains	Proteolytic activation	Serine protease		Cell surface receptor
				activity	substrates	
Apolipoprotein(a)[1]	38	–	?	?	?	?
Prothrombin[2]	2	1 Gla domain	+	+	fibrinogen coagulation factors	?
Factor XII[3]	1	2 EGF like 1 FN type I 1 FN type II	+	+	prekallikrein, factor XI	?
HGF activator[4]	1	2 EGF like 1 FN type I 1 FN type II	+	+	HGF/SF	?
uPA[5]	1	1 EGF like	+	+	plasminogen ECM, HGF/SF	uPA R[10]
tPA[6]	2	1 EGF like 1 FN type I	+	+	plasminogen HGF/SF	low affinity[11]
Plasminogen[7]	5	1 PAP	+	+	Fibrin ECM	low affinity[12]
HGF/SF[8]	4	1 PAP	+	–		c-met + low affinity[13]
HGFl/MSP[9]	4	1 PAP	+	–		Ron[14]
?	?	?	?	?		c-sea[15]

ECM = Extracellular matrix; EGF = epidermal growth factor; FN = fibronectin; HGF = hepatocyte growth factor; HGFl = hepatocyte growth factor-like; MSP = macrophage-stimulating protein; PAP = preactivation peptide; SF = scatter factor; tPA = tissue-type plasminogen activator; uPA = urokinase-type plasminogen activator; uPA R = uPA receptor. Only one reference is shown in each case, usually the first one reporting the cloning of the factor, or the identity of the receptor. Other references can be found in the text.

1 McLean et al. [1987].
2 Degen and Davie [1987].
3 Cool et al. [1985].
4 Miyazawa et al. [1993].
5 Verde et al. [1984].
6 Ny et al. [1984].
7 Novokhatny et al. [1984].
8 Nakamura et al. [1989].
9 Han et al. [1991].
10 Roldan et al. [1990].
11 Carroll et al. [1993].
12 Plow et al. [1986].
13 Naldini et al. [1991b].
14 Gaudino et al. [1994]; Wang et al. [1994].
15 Huff et al. [1993].

a receptor for a kringle-containing serine protease and will not be discussed here.

Plasminogen/Plasmin Receptors. Plasminogen, plasmin, uPA and tPA have been shown to interact with several cell surface binding sites. The main role of the binding sites is supposed to be a local concentration, and therefore an increased efficiency, of this protease system at specific sites of the cell surface where localized extracellular proteolysis is required, such as the leading edge of migrating cells. Plasminogen and plasmin have low-affinity, high-capacity binding sites at the surface of various cells, including some also carrying uPA receptors [Plow et al., 1986]; at least some of these sites could be a plasma membrane-bound form of enolase [Miles et al., 1991]. Binding of plasminogen/plasmin to its receptor increases the rate of formation of plasmin, and protects the latter against inactivation by α_2-antiplasmin.

The uPA receptor is a glycosyl phosphatidyl inositol membrane anchored protein, with a high affinity for the EGF-like domain of the uPA precursor as well as the mature protease [reviewed in Blasi, 1993]. Recently, binding sites for the other plasminogen activator, tPA, have been found [Hajjar, 1991; Bu et al., 1992; Carroll et al., 1993]. The exact nature of these binding sites is not yet known, but the physicochemical characteristics of their interaction with tPA are very different from those of uPA with its receptor, suggesting that this new surface receptor is not related to the uPA receptor. These receptors are thought to potentiate the plasminogen/plasmin system by concentrating its components locally at the surface of migrating cells.

HGF/SF and HGFl/MSP Receptors. The other members of the kringle domain serine protease family for which receptors have been described are the two inactive proteases: HGF/SF and HGFl/MSP, whose only biological effects are mediated by binding to their receptors. These receptors are related tyrosine kinases, defining a subfamily of the tyrosine kinase receptor family; they are respectively called c-met (for HGF/SF) [Bottaro et al., 1991; Naldini et al.,

Fig. 1. Schematic structure of mature HGF/SF generated by proteolytic cleavage at amino acid 465. The A-chain (kringle-containing domain) is in black, the B-chain (serine protease-like domain) in grey. K1–4 = Kringles 1–4; PAP = preactivation peptide.

1991a], and ron (for HGFI/MSP) [Gaudino et al., 1994; Wang et al., 1994]. In chick, another member of the c-met family has been described, c-sea, for which no ligand has yet been found [Huff et al., 1993]. It is therefore possible that other molecules related to HGF/SF and HGFI/MSP, i.e. with kringle domains and an inactive serine protease domain, remain to be described.

Another type of receptor, with low affinity and high capacity, has also been described for HGF/SF [Naldini et al., 1991b]. As for the plasminogen receptor, these binding sites could act by retaining HGF/SF in the pericellular space, and perhaps participate in its activation [for review, see Galimi et al., 1993].

Roles of the Active Serine Proteases during EMT in Development

Circulating Blood Proteases

In addition to its activating properties on platelets, thrombin is chemotactic for monocytes [Bar-Shavit et al., 1983], mitogenic for fibroblasts and vascular smooth muscle cells

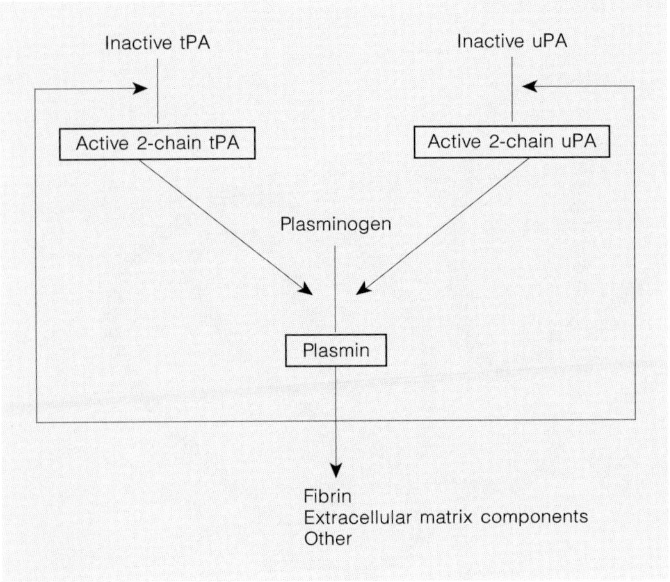

Fig. 2. Schematic cycle leading to plasminogen activation. Arrows indicate proteolytic processing. tPA = Tissue-type plasminogen activator; uPA = urokinase-type plasminogen activator.

[Chen and Buchanan, 1975; McNamara et al., 1992], and strongly inhibits neurite outgrowth in vitro [Suidan et al., 1992], suggesting possible roles during development. Consistent with this hypothesis, expression of the thrombin receptor, a ligand and substrate for thrombin, has been shown in various epithelia and developing organs from 9.5 days p.c. in mouse embryos [Soifer et al., 1994]. However, prothrombin mRNA expression could not be detected before 12.5 days p.c. in mouse [Soifer et al., 1994] and even later in rat [Jamison and Degen, 1991] and is localized exclusively in the developing liver. This observation raises the possibility that another ligand for the thrombin receptor might exist and be involved in its functions during earlier stages of development.

Therefore, there is as yet no evidence that prothrombin is involved in EMT during early development. This result also suggests that the other blood kringle-containing serine protease-like molecules, apolipoprotein(a) and factor XII, are unlikely to play other roles during development. However, no study has yet been published either supporting or conflicting with this hypothesis.

The Plasminogen/Plasmin System

Plasminogen is available to the chick embryo throughout its development, being first captured from the yolk before expression by the embryo proper [Valinsky and Reich, 1981]. In the early phases of mammalian development, it is abundant in the lumen of the uterus [Fazleabas et al., 1983; Finlay et al., 1983]. Its active form, plasmin, is a very pleiotropic protease (see above, among its substrates are extracellular matrix proteins), whose formation must be very tightly regulated throughout development. Several studies have therefore investigated the presence and activity of the two plasminogen activators, uPA and tPA, during development. The importance of these molecules at different stages of embryonic development involving tissue invasion, EMT and cell migration is now well documented.

In one obvious case of developmental EMT, however, plasminogen activators have been reported not to be involved: formation of the primitive streak in the chick [Sanders and Prasad, 1989]. However, this study was performed at a relatively late stage (stage 3–5), when primitive streak formation had already begun. It would therefore be interesting to confirm the results obtained at an earlier stage of chick development. Since this study, nothing has been published concerning the possible roles of kringle-containing serine proteases in vertebrate gastrulation, and such a study is therefore overdue.

Trophoblast Invasion of the Maternal Uterine Mucosa. In mammalian development, implantation of the young embryo in the uterine wall via an EMT-like process implies the localized destruction of the host tissue by the embryo [see Vicovac and Aplin, this vol.]. The cells that initiate this process are the outer cell layer (trophoblast) of the young blastocyst. The involvement of plasminogen activator in this process has been described long ago [Littlefield, 1991]: plasminogen activator inhibitors decrease trophoblast outgrowth in vitro [Kubo et al., 1981], and implantation-defective mouse embryos have reduced plasminogen activator activity [Axelrod, 1985]. It seems likely that only uPA is involved here, since this, but not tPA, is expressed by the trophoblast, as shown by in situ hybridization [Sappino et al., 1989]. Earlier in development, the mouse blastocyst exhibits plasminogen activator activity that derives from the binding of maternal tPA by specific cell surface receptors [Carroll et al., 1993]. At these early stages, tPA-induced production of plasmin by the blastocyst would help its migration through the oviduct. However, this activity disappears slightly before implantation, at which time uPA is the only available plasminogen activator.

Neural Development. The formation of a mature nervous system requires extensive cell movements: immature neurons in the cortex and cerebellum migrate within the neural tube from their initial ventricular location to their final adult position; neural crest cells perform an archetypal EMT [Duband et al., 1995] to leave the dorsal neural tube before migrating to various locations in the embryo; finally, once differentiated, neurons extend axons which travel through very long distances to reach their targets. This axonal extension does not involve an EMT and will not be discussed here. The two plasminogen activators have been shown to be important for these different processes.

Cerebellar granule neurons have plasminogen activator activity in vitro [Krystosek and Seeds, 1981], and express the tPA mRNA in vivo at the time of their migration [Seeds et al., 1992]. Furthermore, protease inhibitors, but not the specific plasmin inhibitor aprotinin, inhibit their migration along neuritic fascicles both in vitro and in vivo [Seeds et al., 1990]. It is therefore likely that tPA is directly involved in the transient detachment of the granule cell from its guiding neurite that occurs during migration.

Similar observations have been made for the migrating neural crest cell population: plasminogen activator activity has been demonstrated for cranial neural crest cells in vivo [Valinsky and Le Douarin, 1985] and trunk crest in vitro [Erickson and Isserroff, 1989]. In trunk neural crest cells, this activity is composed of both uPA and tPA [Menoud et al., 1989], and its inhibition by protease inhibitors prevents crest cell migration [Erickson and Isserroff, 1989]. A possible mechanism of action of uPA and tPA in neural crest cell migration involves TGFβ activation. TGFβ is a growth factor secreted as an inactive precursor which can be proteolytically activated by plasmin [Lyons et al., 1990]. Cranial neural crest cells secrete inactive TGFβ in vitro, and can activate it in the presence of plasminogen [Brauer and Yee, 1993]. Active TGFβ has various effects on neural crest biology [Rogers et al., 1992; Delannet and Duband, 1992], and also decreases the plasminogen activator activity of neural crest cells [Brauer and Yee, 1993], suggesting an autoregulatory mechanism of growth factor production in this cell population.

Heart Development. TGFβ has also been involved in the EMT occurring in the developing heart [Markwald et al., this vol.]: its suppression by antisense oligonucleotides in culture inhibits the transformation of endocardium into migrating mesenchymal cells in the atrioventricular canal [Potts et al., 1991]. These cells will give rise to the mitral and tricuspid valves in mammals [Markwald et al., 1977]. Another study has shown elevated uPA expression associated with the migrating mesenchymal cells [McGuire and Alexander, 1993], and inhibition of uPA synthesis or elimination of its receptors from the cell surface by phospholipase C treatment strongly reduces the motile behavior of these cells in vitro [McGuire and Alexander, 1993]. Here again, uPA could therefore play a role in regulating the motility of endocardial cells through its action on both the surrounding extracellular matrix (either direct or through plasminogen activation), and the TGFβ activity of the migrating cells.

In these different examples of developmental EMTs, the protease activity of tPA and uPA seems therefore to play an important role. But molecular forms of uPA devoid of protease activity have also been shown to increase chemokinesis and chemotaxis in vitro [Gudewicz and Gilboa, 1987; Fibbi et al., 1988; Del Rosso et al., 1993]. A recent study showed that this protease-independent effect of urokinase involves the intracellular phosphorylation of cytoskeletal proteins transduced in the target cells by binding of uPA to its receptor [Busso et al., 1994]. This observation raises the possibility that uPA could also play a growth factor-type role in some cell migration events of morphogenesis. Urokinase could then be more similar to the two nonprotease members of the family: HGF/SF and HGFI/MSP.

Role of the Inactive Serine Protease-Like Molecules: HGF/SF and HGFI/MSP

Although they do not have any protease activity, HGF/SF and HGFI/MSP both exhibit biological activities in vitro and patterns of expression in vivo which suggest roles in EMT and other cell movement related processes during development.

In vitro Effects on Cell Migration and EMT

HGF/SF was independently described 10 years ago for its mitogenic properties towards hepatocytes (hepatocyte growth factor) [Nakamura et al., 1984; Russell et al., 1984], and its ability to induce scattering of epithelial cells in vitro (scatter factor) [Stoker and Perryman, 1985; Stoker et al., 1987]. A few years later, those two apparently very different activities were shown to be due to the same cytokine: a new member of the kringle domain-containing serine protease-like family of molecules [Naldini et al., 1991b; Weidner et al., 1991]. Apart from hepatocytes, HGF/SF is mitogenic for various epithelial and endothelial cells [Matsumoto et al., 1991; Bussolino et al., 1992; Naidu et al., 1994]. But, more relevant to our topic of EMT and cell migration, its motogenic properties have been studied in detail.

A canine kidney cell line, MDCK, was initially used as the choice target for assays of scatter factor activity [Stoker and Perryman, 1985; Stoker et al., 1987]: the cohesive aggregates of MDCK cells in culture become scattered and the cells acquire a high degree of motility in the presence of HGF/SF. The same cells, cultured in three-dimensional collagen gels, form spherical cysts, unless HGF/SF is added to the culture: they then form branched epithelial tubules resembling those formed by the epithelial ureteric duct during kidney morphogenesis in vivo [Montesano et al., 1991; Santos and Nigam, 1993; Santos et al., 1994; see also Davies, this vol.]. These motogenic properties of HGF/SF could be at least partly mediated by an induction of uPA and its receptor in the MDCK cells [Pepper et al., 1992], and they are highly dependent on the extracellular matrix molecules surrounding the cells [Santos and Nigam, 1993; Clark, 1994]. A motogenic effect of HGF/SF has also been shown on endothelial cells [Bussolino et al., 1992] and keratinocytes [Matsumoto et al., 1991].

On the other hand, creation of an autocrine loop by transfection of the HGF/SF receptor, the protooncogene c-met, in cells producing HGF/SF but not normally responding to it, such as NIH3T3 fibroblasts [Giordano et al., 1993; Rong et al., 1994], or conversely, transfection of HGF/SF in HGF/SF responding cells such as MDCK cells [Uehara and Kitamura, 1992] or an epithelial carcinoma cell line [Bellusci et al., 1994], provokes major alterations of the growth and motility of these cells, which might lead to invasiveness and carcinogenesis. Unexpectedly, since it is opposed to the HGF/SF-induced scattering of MDCK cells, which undergo an epithelial-to-mesenchymal transformation, transfection of c-met in HGF/SF producing NIH3T3 fibroblasts was accompanied by the reverse mesenchymal-to-epithelial transformation [Tsarfaty et al., 1994]. Mesenchymal-to-epithelial transformation induced by HGF/SF has also been described for metanephric cell lines [Karp et al., 1994]. These various effects suggest that HGF/SF could be important in the processes of EMT and transitions in the reverse direction (METs) occurring during the formation of various organs (such as the kidney, see Davies [this vol.]; or cancer invasion in the early stages of metastasis, see Birchmeier et al. [this vol.]).

The biological effects of HGFl/MSP, on the other hand, are much less well understood. Cloned because of its sequence homology to prothrombin, and identified as a new molecule sharing the structural features of HGF/SF [Degen et al., 1991; Han et al., 1991], it had no known biological activity until its identity with another cytokine, MSP, was discovered [Bezerra et al., 1993; Shimamoto et al., 1993; Yoshimura et al., 1993]. HGFl/MSP induces in vitro the responsiveness of peritoneal macrophages to chemoattractants, and increases their complement-induced phagocytic properties [Leonard and Skeel, 1976; Skeel et al., 1991]. HGFl/MSP also affects directly the morphology and motility of resident macrophages in vitro, for which it is a chemoattractant [Skeel and Leonard, 1994]. Although this effect has only been shown on one cell type, monocytes, the motogenic activity of HGFl/MSP is consistent with a possible role in cell migration, which is a frequent consequence of EMT.

In vivo Effects during Development

Analysis of patterns of expression in situ, combined with experiments performed in vivo, have provided arguments for roles of HGF/SF and HGFl/MSP during various events in embryonic development.

Early Development: Gastrulation and Neural Tube Formation. A model for morphogenesis of the primitive streak in the chick embryo [Stern, 1984] made the prediction that local loss of epithelial continuity in the epiblast of the early embryo would lead to the formation of a primitive streak at that site and inhibition of the formation of the original embryonic axis. For this reason, and because of the reported scattering effects of HGF/SF on cultured epithelial cells, the effect of its local application to young chick embryos was tested [Stern et al., 1990]. Beads impregnated with HGF/SF or cells expressing it grafted in the gastrulating chick embryo were indeed found to cause partial axial duplications, but unexpectedly, these included the formation of a neural plate like structure in some cases [Stern et al., 1990]. This led to the suggestion that HGF/SF may be involved more directly in neural induction. Therefore, an in vitro assay was designed to test this. HGF/SF was found to induce strong expression of the early neural marker, L5, and to promote neural differentiation [Streit et al., 1995]. Because antibodies to L5 interfere with neural induction in vivo, and because L5 is initially expressed in cells that are competent to respond to neural induction by the organizer (Hensen's node), it was proposed that HGF/SF may be involved in the early steps of this process, perhaps by inducing or maintaining neural competence in ectodermal cells.

Further evidence for this proposal was obtained after cloning the chick homologs of HGF/SF, HGFl/MSP and the receptor c-met, and studying their expression by in situ hybridization and RNAase protection [Streit et al., 1995; Théry et al., 1995]. The first was found to be expressed in Hensen's node but downregulated at precisely the stage when this structure loses its neural-inducing ability [Streit et al., 1995], whereas its receptor seems to be expressed in the whole embryo at this stage, albeit at a very low level, as shown by RNAase protection assays [Théry et al., 1995]. In the early mouse embryo, HGF/SF is expressed in the anterior part of the primitive streak together with c-met [Andermarcher et al., submitted].

At the same time as gastrulation and neural induction are taking place, HGFl/MSP is expressed in the whole of the embryonic region (area pellucida) but not in the future extraembryonic areas. Subsequently, it becomes restricted first to the neural plate, then to a portion of the notochord and finally to the floor plate region of the spinal cord [Théry et al., 1995]. Expression in the latter two structures is closely coordinated, such that it is downregulated from the notochord at a similar axial level as it appears in the floor plate. This suggests that HGFI/MSP could play a role in induction of the floor plate, perhaps acting together with *Shh/Vhh-1* [Riddle et al., 1993; Roelink et al., 1994]. In the frog too, a member of the HGF/SF-HGFI/MSP family is expressed at these stages of embryogenesis, in regions including the neural plate [Ruiz i Altaba, pers. commun.].

Somite Morphogenesis. Both HGF/SF and HGFI/MSP, as well as the receptor c-met are expressed in complementary portions of the developing somite mesoderm of the chick embryo. c-met is the first of these to be expressed, in the first 3–4 somites to form, which will later contribute to the occipital bone and to some craniofacial muscles. After this, c-met mRNA is not detectable until about the 13-somite stage, when a narrow band of expression appears at the dorsomedial edge of each somite, corresponding to the dorsal edge of the dermomyotome [Théry et al., 1995]. HGF/SF is expressed only weakly in somites, but appears to be restricted to the sclerotome portion (which will contribute to the axial skeleton); HGFI/MSP, on the other hand, is expressed very strongly and is restricted to the myotome (which will give rise to the axial musculature) [Théry et al., 1995]. These findings implicate all three macromolecules in the development of somites and their derivatives, although at present their precise roles are unclear. It is interesting to note, however, that several components of the somite mesoderm undergo EMT or the reverse transformation (MET) during their development: first, the mesenchymal unsegmented mesoderm becomes an epithelial sphere, the newly-formed somite. This structure remains epithelial until about 10 h later, when its ventromedial part loses its epithelial structure and becomes mesenchymal once again, to form the sclerotome (which expresses HGF/SF). The remaining, dorsolateral, part of the somite remains epithelial for longer, as the dermomyotome, but then its dorsomedial edge (which expresses c-met) involutes to form the myotome (which expresses HGFI/MSP).

Organogenesis and Limb Development. HGF/SF, HGFI/MSP and c-met are also expressed in various organ primordia during their development. They include the developing heart [Andermarcher et al., submitted; Théry et al., 1995], the liver and kidney [Andermarcher et al., submitted; Sonnenberg et al., 1993]. At these later stages of development, it seems that c-met tends to be expressed in the epithelial component and HGF/SF in the adjacent mesenchyme [Andermarcher et al., submitted; Sonnenberg et al., 1993]. Each of these organ primordia also contains cells that undergo EMT. The heart includes cells derived from the endothelial lining which become mesenchymal and migratory and contribute to the endocardial cushions and valves [see above and Markwald et al., this vol.]. The liver is of mixed endodermal and mesodermal origin and not much is known about its development, but it must include the formation of epithelial ducts and hepatic diverticula as well as the organization of the looser cells into the mesenchymal surrounding of the portal triads. The adult (metanephric) kidney is also derived from both mesenchymal and epithelial cells. Some of the mesenchymal cells condense to form the kidney tubules; as discussed above and by Davies [this vol.], a tubulogenic effect of HGF/SF on MDCK cells has been reported in vitro [Montesano et al., 1991; Santos and Nigam, 1993; Santos et al., 1994].

Finally, the strong expression of HGF/SF in the developing limb buds is worthy of mention. This is seen as soon as the limb buds of the chick embryo first become discernible [Théry et al., 1995] and it is therefore possible that this factor could be involved in determining the regions of the flank that undergo limb bud elevation, perhaps acting in concert with members of the FGF family of molecules [Cohn et al., 1995].

Angiogenesis. During embryonic development, expression of HGF/SF is observed in the forming blood islands in the extraembryonic area vasculosa of the chick embryo, as well as in the endothelial lining of the aortic arches [Théry et al., 1995]. This observation is consistent with the known mitogenic and motogenic effects on cultured endothelial cells [Bussolino et al., 1992] as well as with the finding that a gel fragment impregnated with HGF/SF, implanted under the skin of a mouse strongly promotes angiogenesis in the vicinity of the implant [Grant et al., 1993]. These findings implicate HGF/SF in the process of blood vessel formation in vivo, both in the embryo and in the adult; it was therefore proposed that deregulation of this effect could underlie certain pathologies such as psoriasis [Grant et al., 1993] or AIDS-related Kaposi sarcoma [Naidu et al., 1994].

HGF/SF-Deficient Mice

Two recent publications report the production of homozygous null mutations in the gene encoding HGF/SF in mice by homologous recombination [Schmidt et al., 1995; Uehara et al., 1995]. In both reports, the mutation is found to be lethal at stages of early organogenesis, but the reported cause of death differs between the two studies. In one [Schmidt et al., 1995] the mutant animals die at about day 13–13.5, apparently because of liver defects; in the other [Uehara et al., 1995] the mice are found to be dead from day

12.5, apparently because of defects in the placental circulation and the formation of placental lacunae. The two findings appear difficult to reconcile at present, but it is possible that differences in the constructs used by the two groups, which placed the *neo* insertion into the fifth and second exons, respectively, are responsible. If so, the difference in phenotype suggests that two different regions situated close to each other in the amino-terminal part of the protein serve different functions and perhaps bind to different receptors.

It is also interesting, but perhaps not surprising, that neither group reported phenotypic effects of the mutation on the development in any other organ system or tissue in which HGF/SF or c-met is expressed at earlier stages of development. This is an increasingly common finding using mutant mice generated by homologous recombination, and stresses the remarkable ability of developing systems to compensate for a variety of insults ('regulation').

Conclusion

In conclusion, several observations suggest that HGF/SF, HGFl/MSP and the c-met receptor are involved in different EMT as well as other events during embryonic development. To complete these studies, analysis of expression of Ron, the HGFl/MSP receptor, during development will be very informative. However, apart from HGF/SF in neural induction [Streit et al., 1995], functional experiments demonstrating the roles of these molecules are still lacking. Since both HGF/SF and HGFl/MSP are secreted as inactive precursors, which need to be proteolytically cleaved to give rise to the active molecule, it is also desirable to study the regulation of this activation, for instance by analyzing the expression pattern of the HGF/SF activator [Miyazawa et al., 1993] during embryonic development. Finally, the existence of a growth factor of the HGF/SF-HGFl/MSP family in *Xenopus* [Ruiz i Altaba, pers. commun.], as well as of a receptor related to c-met and Ron in chick [Huff et al., 1993], but still orphan, suggests that the family of kringle-containing serine-protease like growth factors might contain additional members with potential roles in development.

Acknowledgements

Our research on this subject was funded by the Muscular Dystrophy Association. We are grateful to Dr. Andrea Streit for comments on the manuscript.

References

Andermarcher, E., M.A. Surani, E. Gherardi (submitted) Distinct patterns of expression of the HGF/SF and c-met genes during early and late mouse embryogenesis.

Axelrod, H.R. (1985) Altered trophoblast functions in implantation-defective mouse embryos. Dev Biol *108:* 185–190.

Bar-Shavit, R., A. Kahn, G. Wilner, J.I. Fenton (1983) Monocyte chemotaxis: stimulation by specific exosite region in thrombin. Science *220:* 728–731.

Bellusci, S., G. Moens, G. Gaudino, P. Comoglio, T. Nakamura, J.-P. Thiery, J. Jouanneau (1994) Creation of an hepatocyte growth factor/scatter factor autocrine loop in carcinoma cells induces invasive properties associated with increased tumorigenicity. Oncogene *9:* 1091–1099.

Bezerra, J.A., S. Han, M.J.S. Danton, S.J.F. Degen (1993) Are hepatocyte growth factor-like protein and macrophage stimulating protein the same protein? Protein Sci *2:* 666–668.

Blasi, F. (1993) Urokinase and urokinase receptor: A paracrine/autocrine system regulating cell migration and invasiveness. BioEssays *15:* 105–111.

Bottaro, D.P., J.S. Rubin, D.L. Faletto, A.M.-L. Chan, T.E. Kmiecick, G.F. Vande Woude, S.A. Aaronson (1991) Identification of the hepatocyte growth factor receptor as the c-met proto-oncogene product. Science *251:* 802–804.

Brauer, P.R., J.A. Yee (1993) Cranial neural crest cells synthesize and secrete a latent form of transforming growth factor β that can be activated by neural crest cell proteolysis. Dev Biol *155:* 281–285.

Bu, G., S. Williams, D.K. Strickland, A.L. Schwartz (1992) Identification and partial characterization by chemical cross-linking of a binding protein for tissue-type plasminogen activator on rat hepatoma cells. J Biol Chem *267:* 15595–15602.

Busso, N., S.K. Masur, D. Lazega, S. Waxman, L. Ossowski (1994) Induction of cell migration by pro-urokinase binding to its receptor: Possible mechanism for signal transduction in human epithelial cells. J Cell Biol *126:* 259–270.

Bussolino, F., M.F. Di Renzo, M. Ziche, E. Bocchietto, M. Olivero, L. Naldini, G. Gaudino, L. Tamagnone, A. Coffer, P.M. Comoglio (1992) Hepatocyte growth factor is a potent angiogenic factor which stimulates endothelial cell motility and growth. J Cell Biol *119:* 629–641.

Carroll, P.M., W.G. Richards, A.L. Darrow, J.M. Wells, S. Strickland (1993) Preimplantation mouse embryos express a cell surface receptor for tissue-plasminogen activator. Development *119:* 191–198.

Chen, L., J. Buchanan (1975) Mitogenic activity of blood components: Thrombin and prothrombin. Proc Natl Acad Sci USA *72:* 1321–1325.

Clark, P. (1994) Modulation of scatter factor/hepatocyte growth factor activity by cell substratum adhesion. J Cell Sci *107:* 1265–1275.

Cohn, M.J., J.-C. Izpisùa-Belmonte, C. Tickle (1995) FGF2 application can induce additional limb formation from the flank of the chick embryo. Cell *80:* 739–746.

Cool, D.E., C.-J.S. Edgell, G.V. Louie, M.J. Zoller, G.D. Brayer, R.T.A. MacGillivray (1985) Characterization of human blood coagulation factor XII cDNA: Prediction of the primary structure of factor XII and the tertiary structure of β-factor XIIa. J Biol Chem *260:* 13666–13676.

Degen, S.J.F., E.W. Davie (1987) Nucleotide sequence of the gene for human prothrombin. Biochemistry *26:* 6165–6177.

Degen, S.J.F., L.A. Stuart, S. Han, C.S. Jamison (1991) Characterization of the mouse cDNA and gene coding for a hepatocyte growth factor-like protein: Expression during development. Biochemistry *30:* 9781–9791.
Delannet, M., J.-L. Duband (1992) Transforming growth factor β control of cell-substratum adhesion during avian neural crest cell emigration in vitro. Development *116:* 275–287.
Del Rosso, M., E. Anichini, N. Pedersen, F. Blasi, G. Fibbi, M. Pucci, M. Ruggiero (1993) Urokinase-urokinase receptor interaction: Nonmitogenic signal transduction in human epidermal cells. Biochem Biophys Res Commun *190:* 347–352.
Duband, J.-L., F. Monier, M. Delannet, D. Newgreen (1995) Epithelium-mesenchyme transition during neural crest development. Acta Anat *154:* 63–78.
Erickson, C.A., R.R. Isserroff (1989) Plasminogen activator activity is associated with neural crest cell motility in tissue culture. J Exp Zool *251:* 123–133.
Fazleabas, A.T., R.D. Geisert, P.W. Bazer, R.M. Roberts (1983) Relationship between release of plasminogen activator and estrogen by blastocysts and secretion of plasmin inhibitor by uterine endometrium in the pregnant pig. Biol Reprod *29:* 225–238.
Fibbi, G., M. Ziche, L. Morbidelli, L. Magnelli, M. Del Rosso (1988) Interaction of urokinase with specific receptors stimulates mobilization of bovine adrenal capillary endothelial cells. Exp Cell Res *179:* 385–395.
Finlay, T.H., J. Katz, L. Kirsch, M. Levitz, S.A. Nathoo, S. Seiler (1983) Estrogen-stimulated uptake of plasminogen by the mouse uterus. Endocrinology *112:* 856–861.
Galimi, F., M.F. Brizzi, P.M. Comoglio (1993) The hepatocyte growth factor and its receptor. Stem Cell *11:* 22–30.
Gaudino, G., A. Follenzi, L. Naldini, C. Collesi, M. Santoro, K.A. Gallo, P.J. Godowski, P.M. Comoglio (1994) Ron is a heterodimeric tyrosine kinase receptor activated by the HGF homologue MSP. EMBO J *13:* 3524–3532.
Giordano, S., Z. Zhen, E. Medico, G. Gaudino, F. Galimi, P.M. Comoglio (1993) Transfer of motogenic and invasive response to scatter factor/hepatocyte growth factor by transfection of human MET protooncogene. Proc Natl Acad Sci USA *90:* 649–653.
Grant, D.S., H.K. Kleinman, I.D. Goldberg, M.M. Bhargava, B.J. Nickoloff, J.L. Kinsella, P. Polverini, E.M. Rosen (1993) Scatter factor induces blood vessel formation in vivo. Proc Natl Acad Sci USA *90:* 1937–1941.
Gudewicz, P.W., N. Gilboa (1987) Human urokinase-type plasminogen activator stimulates chemotaxis of human neutrophils. Biochem Biophys Res Commun *147:* 1176–1181.
Hajjar, K.A. (1991) The endothelial cell tissue plasminogen activator. J Biol Chem *266:* 21962–21970.
Han, S., L.A. Stuart, S.J.F. Degen (1991) Characterization of the DNF1582 locus on human chromosome 3: Identification of a gene coding for four kringle domains with homology to hepatocyte growth factor. Biochemistry *30:* 9768–9780.
Hay, E.D. (1995) An overview of epithelio-mesenchymal transformation. Acta Anat *154:* 8–20.
Huff, J., M.A. Jelinek, C.A. Borgman, T.J. Lansing, J.T. Parsons (1993) The protoncogene c-sea encodes a transmembrane receptor protein-tyrosine kinase related to the Met/hepatocyte growth factor/scatter factor receptor. Proc Natl Acad Sci USA *90:* 6140–6144.
Jamison, C.S., S.J.F. Degen (1991) Prenatal and postnatal expression of mRNA coding for rat prothrombin. Biochim Biophys Acta *1088:* 208–216.
Karp, S.L., A. Ortiz-Arduan, S. Li, E.G. Neilson (1994) Epithelial differentiation of metanephric mesenchymal cells after stimulation with hepatocyte growth factor or embryonic spinal cord. Proc Natl Acad Sci USA *91:* 5286–5290.
Krystosek, A., N.W. Seeds (1981) Plasminogen activator secretion by granule neurons in cultures of developing cerebellum. Proc Natl Acad Sci USA *78:* 7810–7814.
Kubo, H., A. Spindle, R.A. Pedersen (1981) Inhibition of mouse blastocyst attachment and outgrowth by protease inhibitors. J Exp Zool *216:* 445–451.
Leonard, E.J., A. Skeel (1976) A serum protein that stimulates macrophage movement, chemotaxis and spreading. Exp Cell Res *102:* 434–438.
Littlefield, B.A. (1991) Plasminogen activators in endometrial physiology and embryo implantation: A review. Ann NY Acad Sci *622:* 167–175.
Lyons, R.M., L.E. Gentry, A.F. Purchio, H.L. Moses (1990) Mechanism of activation of latent recombinant transforming growth factor β1 by plasmin. J Cell Biol *110:* 1361–1367.
Markwald, R.R., T.P. Fitzharris, F.J. Manasek (1977) Structural analysis of endocardial cushion development. Am J Anat *148:* 85–120.
Matsumoto, K., K. Hashimoto, K. Yoshikawa, T. Nakamura (1991) Marked stimulation of growth and motility of human keratinocytes by hepatocyte growth factor. Exp Cell Res *196:* 114–120.
McGuire, P.G., S.M. Alexander (1993) Inhibition of urokinase synthesis and cell surface binding alters the motile behavior of embryonic endocardial-derived mesenchymal cells in vitro. Development *118:* 931–939.
McGuire, P.G., N.W. Seeds (1990) Degradation of underlying extracellular matrix by sensory neurons during neurite outgrowth. Neuron *4:* 633–642.
McLean, J.W., J.E. Tomlinson, W.-J. Kuang, D.L. Eaton, E.Y. Chen, G.M. Fless, A.M. Scanu, R.M. Lawn (1987) cDNA sequence of human apolipoprotein(a) is homologous to plasminogen. Nature *330:* 132–137.
McNamara, C., I. Sarembok, L. Gimple, J.W. Fenton II, S.R. Coughlin, G. Owens (1992) Thrombin stimulation of smooth muscle cell proliferation is mediated by a proteolytic receptor-mediated mechanism. J Clin Invest *91:* 94–98.
Menoud, P.A., S. Debrot, J. Schowing (1989) Mouse neural crest cells secrete both urokinase-type and tissue-type plasminogen activators in vitro. Development *106:* 685–690.
Miles, L.A., C.M. Dahlberg, J. Plescia, J. Felez, K. Kato, F. Plow (1991) Role of cell-surface lysines in plasminogen binding to cells: Identification of alpha-enolase as a candidate plasminogen receptor. Biochemistry *30:* 1682–1691.
Miyazawa, K., T. Shimomura, A. Kitamura, J. Kondo, Y. Morimoto, N. Kitamura (1993) Molecular cloning and sequence analysis of the cDNA for a human serine protease responsible for activation of hepatocyte growth factor: Structural similarity of the protease precursor to blood coagulation factor XII. J Biol Chem *268:* 10024–10028.
Montesano, R., K. Matsumoto, T. Nakamura, L. Orci (1991) Identification of a fibroblast-derived epithelial morphogen as hepatocyte growth factor. Cell *67:* 901–908.
Naidu, Y.M., E.M. Rosen, R. Zitnick, I. Goldberg, M. Park, M. Naujokas, P.J. Polverini, B.J. Nickoloff (1994) Role of scatter factor in the pathogenesis of AIDS-related Kaposi sarcoma. Proc Natl Acad Sci USA *91:* 5281–5285.
Nakamura, T., K. Nawa, A. Ichihara (1984) Partial purification and characterization of a hepatocyte growth factor from serum of hepatectomized rat. Biochem Biophys Res Commun *122:* 1450–1459.
Nakamura, T., T. Nishizawa, M. Hagiya, T. Seki, M. Shimonishi, A. Sugimara, K. Tashiro, S. Shimizu (1989) Molecular cloning and expression of human hepatocyte growth factor. Nature *342:* 440–443.
Naldini, L., E. Vigna, R.P. Narsimhan, G. Gaudino, R. Zarnegar, G.K. Michalopoulos, P.M. Comoglio (1991a) Hepatocyte growth factor (HGF) stimulates the tyrosine kinase activity of the receptor encoded by the proto-oncogene c-met. Oncogene *6:* 501–504.
Naldini, L., K.M. Weidner, E. Vigna, G. Gaudino, A. Bardelli, C. Ponzetto, R.P. Narsimhan, G. Hartmann, R. Zarnegar, G.K. Michalopoulos, W. Birchmeier, P. Comoglio (1991b) Scatter factor and hepatocyte growth factor are indistinguishable ligands for the MET receptor. EMBO J *10:* 2867–2878.
Novokhatny, V.V., S.A. Kudinov, P.L. Privalov (1984) Domains in human plasminogen. J Mol Biol *179:* 215–232.
Ny, T., F. Elgh, B. Lund (1984) The structure of the human tissue-type plasminogen activator gene: correlation of intron and exon structures to functional and structural domains. Proc Natl Acad Sci USA *81:* 5355–5359.
Pepper, M.S., K. Matsumoto, T. Nakamura, L. Orci, R. Montesano (1992) Hepatocyte growth factor increases urokinase-type plasminogen activator (u-PA) and u-PA receptor expression in Madin-Darby canine kidney epithelial cells. J Biol Chem *267:* 20493–20496.
Plow, E.F., D.E. Freaney, J. Plescia, L.A. Miles (1986) The plasminogen system and cell surfaces: Evidence for plasminogen and urokinase receptors on the same cell type. J Cell Biol *103:* 1411–1420.

Potts, J.D., J.M. Dagle, J.A. Walder, D.L. Weeks, R.B. Runyan (1991) Epithelial-mesenchymal transformation of embryonic cardiac endothelial cells is inhibited by a modified antisense oligodeoxynucleotide to transforming growth factor β3. Proc Natl Acad Sci USA *88:* 1516–1520.

Quigley, J.P., L. Gold, R. Schwimmer, L. Sullivan (1987) Limited cleavage of cellular fibronectin by plasminogen activator purified from transformed cells. Proc Natl Acad Sci USA *84:* 2776–2780.

Riddle, R.D., R.L. Johnson, E. Laufer, C.J. Tabin (1993) Sonic hedgehog mediates the polarizing activity of the ZPA. Cell *75:* 1401–1418.

Roelink, H., A. Augsburger, J. Heemskerk, V. Korzh, S. Norlin, A. Ruiz i Altaba, Y. Tanabe, M. Placzek, T. Edlund, T.M. Jessell, T. Dodd (1994) Floor plate and motor neuron induction by vhh-1, a vertebrate homologue of hedgehog expressed by the notochord. Cell *76:* 761–775.

Rogers, S.L., P.J. Gegick, S.M. Alexander, P.G. McGuire (1992) Transforming growth factor-β alters differentiation in cultures of avian neural crest-derived cells: Effects on cell morphology, proliferation, fibronectin expression, and melanogenesis. Dev Biol *151:* 192–203.

Roldan, A.L., M.V. Cubellis, M.T. Masucci, N. Behrendt, L.R. Lund, K. Danø, E. Appella, F. Blasi (1990) Cloning and expression of the receptor for human urokinase plasminogen activator, a central molecule in cell surface, plasmin dependent proteolysis. EMBO J *9:* 467–474.

Rong, S., S. Segal, M. Anver, J.H. Resau, G.F. Vande Woude (1994) Invasiveness and metastasis of NIH 3T3 cells induced by Met-hepatocyte growth factor/scatter factor autocrine stimulation. Proc Natl Acad Sci USA *91:* 4731–4735.

Russell, W.E., J.A. McGowan, N.R.L. Bucher (1984) Partial characterization of a hepatocyte growth factor from rat platelets. J Cell Physiol *119:* 183–192.

Sanders, E.J., S. Prasad (1989) Evidence that plasminogen activator is not involved in basement membrane penetration at avian gastrulation. Exp Cell Res *185:* 394–398.

Santos, O.F.P., E.J.G. Barros, X.-M. Yang, K. Matsumoto, T. Nakamura, M. Park, S.K. Nigam (1994) Involvement of hepatocyte growth factor in kidney development. Dev Biol *163:* 525–529.

Santos, O.F.P., S.K. Nigam (1993) HGF-induced tubulogenesis and branching of epithelial cells is modulated by extracellular matrix and TGFβ. Dev Biol *160:* 293–302.

Sappino, A.-P., J. Huarte, D. Belin, J.-D. Vassalli (1989) Plasminogen activators in tissue remodeling and invasion: mRNA localization in mouse ovaries and implanting embryos. J Cell Biol *109:* 2471–2479.

Schmidt, C., F. Bladt, S. Goedecke, V. Brinkmann, W. Zschiesche, M. Sharpe, E. Gherardi, C. Birchmeier (1995) Scatter factor/hepatocyte growth factor is essential for liver development. Nature *373:* 699–702.

Seeds, N.W., S. Haffke, K. Christensen, J. Schoomaker (1990) Cerebellar granule cell migration involves proteolysis. Adv Exp Med Biol *265:* 169–178.

Seeds, N.W., S. Verrall, G., Friedman, S. Hayden, D. Gadotti, S. Haffke, K. Christensen, B. Gardner, P. McGuire, A. Krystosek (1992) Plasminogen activators and plasminogen activator inhibitors in neural development. Ann NY Acad Sci *667:* 32–40.

Shimamoto, A., T. Kimura, K. Matsumoto, T. Nakamura (1993) Hepatocyte growth factor-like protein is identical to macrophage stimulating protein. FEBS Lett *333:* 61–64.

Skeel, A., E.J. Leonard (1994) Action and target cell specificity of human macrophage-stimulating protein (MSP). J Immunol *152:* 4618–4623.

Skeel, A., T. Yoshimura, S.D. Showaltzer, S. Tanaka, E. Appella, E.J. Leonard (1991) Macrophage stimulating protein: Purification, partial amino acid sequence, and cellular activity. J Exp Med *173:* 1227–1234.

Soifer, S.J., K.G. Peters, J. O'Keefe, S.R. Coughlin (1994) Disparate temporal expression of the prothrombin and thrombin receptor genes during mouse development. Am J Pathol *144:* 60–69.

Sonnenberg, E., D. Meyer, K.M. Weidner, C. Birchmeier (1993) Scatter factor/hepatocyte growth factor and its receptor, the c-met tyrosine kinase, can mediate a signal exchange between mesenchyme and epithelia during mouse development. J Cell Biol *123:* 223–235.

Stern, C.D. (1984) A simple model for early morphogenesis. J Theor Biol *107:* 229–242.

Stern, C.D., G.W. Ireland, S.E. Herrick, E. Gherardi, J. Gray, M. Perryman, M. Stoker (1990) Epithelial scatter factor and development of the chick embryonic axis. Development *110:* 1271–1284.

Stoker, M., E. Gherardi, M. Perryman, J. Gray (1987) Scatter factor is a fibroblast-derived modulator of epithelial cell mobility. Nature *327:* 239–242.

Stoker, M., M. Perryman (1985) An epithelial scatter factor released by embryo fibroblasts. J Cell Sci *77:* 209–213.

Streit, A., C.D. Stern, C. Théry, G. Ireland, S. Aparicio, M.J. Sharpe, E. Gherardi (1995) A role for HGF/SF in neural induction and its expression in Hensen's node during gastrulation. Development *121:* 813–824.

Suidan, H.S., S.R. Stone, B.A. Hemmings, D. Monard (1992) Thrombin causes neurite retraction in neuronal cells through activation of cell surface receptors. Neuron *8:* 363–375.

Théry, C., M.J. Sharpe, S.J. Batley, C.D. Stern, E. Gherardi (1995) Expression of HGF/SF, HGFl/MSP and c-met suggests new functions during early chick development. Dev Genet *17:* 90–101.

Tsarfaty, I., S. Rong, J.H. Resau, S. Rulong, P. Pinto da Silva, G.F. Vande Woude (1994) The Met proto-oncogene mesenchymal to epithelial cell conversion. Science *263:* 98–101.

Uehara, Y., N. Kitamura (1992) Expression of a human hepatocyte growth factor/scatter factor cDNA in MDCK epithelial cells influences cell morphology, motility, and anchorage-independent growth. J Cell Biol *117:* 889–894.

Uehara, Y., O. Minowa, C. Mori, K. Shiota, J. Kuno, T. Noda, N. Kitamura (1995) Placental defect and embryonic lethality in mice lacking hepatocyte growth factor/scatter factor. Nature *373:* 702–705.

Valinsky, J.E., N.M. Le Douarin (1985) Production of plasminogen activator by migrating cephalic neural crest cells. EMBO J *4:* 1403–1406.

Valinsky, J.E., E. Reich (1981) Plasminogen in the chick embryo: Transport and biosynthesis. J Biol Chem *256:* 12470–12475.

Vassalli, J.-D., A.-P. Sappino, D. Belin (1991) The plasminogen activator/plasmin system. J Clin Invest *88:* 1067–1072.

Verde, P., M.P. Stoppelli, P. Galeffi, P. Di Nocera, F. Blasi (1984) Identification and primary sequence of an unspliced human urokinase poly(A)+RNA. Proc Natl Acad Sci USA *81:* 4727–4731.

Vu, T.-K.H., D.T. Hung, V.I. Wheaton, S.R. Coughlin (1991) Molecular cloning of a functional thrombin receptor reveals a novel proteolytic mechanism of receptor activation. Cell *64:* 1057–1068.

Wang, M.-H., C. Ronsin, M.-C. Gesnel, L. Coupey, A. Skeel, E.J. Leonard, R. Breathnach (1994) Identification of the Ron gene product as the receptor for the human macrophage stimulating protein. Science *266:* 117–119.

Weidner, K.M., N. Arakaki, G. Hartmann, J. Vanderkerckhove, S. Weingart, H. Rieder, C. Fonatsch, H. Tsubouchi, T. Hishida, Y. Daikuhara, W. Birchmeier (1991) Evidence for the identity of human scatter factor and human hepatocyte growth factor. Proc Natl Acad Sci USA *88:* 7001–7005.

Yoshimura, T., N. Yuhki, M.-H. Wang, A. Skeel, E.J. Leonard (1993) Cloning, sequencing and expression of human macrophage stimulating protein (MSP, MST1) confirms MSP as a member of the family of kringle proteins and locates the MSP gene on chromosome 3. J Biol Chem *268:* 15461–15468.

Acta Anatomica

Acta Anat 1996;156:173–186

R. Markwald
C. Eisenberg
L. Eisenberg
T. Trusk
Y. Sugi

Department of Cell Biology and Anatomy, Medical University of South Carolina, Charleston, S.C., USA

Epithelial-Mesenchymal Transformations in Early Avian Heart Development

Key Words
Heart development
Epithelial-mesenchymal transformations
Endothelium
Induction
Extracellular matrix

Abstract
Cardiac morphogenesis proceeds from a sequential series of epithelial-mesenchymal transitions which begins by establishing bipotential heart-forming cells and later their segregation into endocardial and myocardial lineages. Cells within each lineage integrate to form two concentric epithelia which inductively interact to transform cells of the inner epithelium, the endocardium, into mesenchymal or 'cushion' cells. Noncardiogenic epithelia (dorsal mesocardium, epicardium, neural ectoderm and coelomic mesothelium) undergo transition into populations of extracardiac mesenchyme that combine over time with cushion tissue to remodel the simple tubular heart into a four-chambered organ. Model systems are described for studying the mechanisms of cardiac-related transformations including primary cultures of precardiac epithelia and a differentiation-inducible, avian stem cell line called QCE-6. Focus is centered on the molecular mechanism by which endocardial epithelium transforms into cushion mesenchyme. Experimental findings are reviewed and interpreted in the context of a hypothetical model that seeks to answer why only some cells within an epithelium transform and whether the transformation process is regulated by intrinsic or extrinsic mechanisms. The model proposes that epithelial cells competent to transform to mesenchyme express characteristic markers including receptors for extrinsic signals secreted by stimulator cells (e.g. myocardium). Candidate extrinsic signals include multicomponent complexes called adherons. If applied directly to cultured endocardium, myocardial adherons but not those secreted by L6 myoblasts, induce changes in gene expression within target endocardial cells for proteases and cell:cell and cell:matrix adhesion molecules that accompanied transformation to mesenchyme. A main component of myocardial adherons has been identified as ES antigens, one of which, ES/130, has been cloned, found to have a novel sequence and in culture assays shown to be required for endocardium to transform to mesenchyme. The spatiotemporal pattern of ES protein expression within the embryo suggests that common mechanisms may exist for embryonic epithelial-mesenchymal transformations.

Abbreviations used in this paper:
AV = Atrioventricular; CJ = cardiac jelly; ES = EDTA-soluble; FGF = fibroblast growth factor; N-CAM = neural cell adhesion molecule; QCE-6 = quail cardiac cell line; TGF = transforming growth factor.

KARGER
E-Mail karger@karger.ch
Fax + 41 61 306 12 34
http://www.karger.ch

R.R. Markwald
Department of Cell Biology and Anatomy
Medical University of South Carolina
Charleston, SC 29425 (USA)

Introductory Overview of Cardiac Relevant, Epithelial Transformations

Morphogenesis of the heart begins as primary mesodermal cells become committed to the cardiac lineage, and coalesce into a hollow tubular structure. Subsequently, the initial heart tube is remodeled into a beating, multichambered organ. This description of cardiac morphogenesis has become increasingly complex as our understanding of the molecular and cellular events which occur during the development of the heart has increased. Thus, as stated by Icardo [1988], the challenge of understanding heart development is to assimilate this new information into fundamental concepts to explain the origin and integration of multiple cardiac cell types. Accordingly, a morphogenetic event that has been shown to be crucial to cardiac development is the transformation of epithelium into mesenchyme. We envision heart development as proceeding from a series of such transformations that collectively and progressively establish the 3-dimensional patterns of cardiac tissue organization by producing: (i) the precursor cells for the paired heart forming fields; (ii) the concentric epithelia of the primary heart tube (myocardium and endocardium); (iii) coronary vasculature, and (iv) septal structures of intracardiac and extracardiac origin that subdivide the primitive heart tube into four chambers. In table 1, we list six, cardiac-relevant, epithelial transformations. Of these, molecular regulation is best understood for endocardial transformation to cushion mesenchyme. In the present review, we focus on this particular transformation with the view that it might constitute a prototype for the other cardiac relevant, epithelial transformations about which much less is known.

From Gastrulation to the Formation of Paired Heart Fields (fig. 1)

Mesodermal cells specified to heart lineages emerge from the epithelial epiblast by migration through the rostral half of the primitive streak [Garcia-Martinez and Schoenwolf, 1993]. This occurs in proximity to Hensen's node, an early polarizing site that secretes the morphogen, retinoic acid [Chen et al., 1992]. Cells committed to a heart lineage migrate anteriorly and ventrally upon the endodermal basement membrane. Their constitutive expression of an antigen, termed JB3, permits their visualization as paired primordia, the 'heart-forming fields', within the larger, general population of future lateral plate mesodermal cells [Wunsch et al., 1994].

Table 1. Epithelial-mesenchymal transformations in heart development

1	Gastrulation: establishment of the heart-forming fields
2	Heart-forming fields (cardiogenic mesoderm): differentiation of bipotential precardiac stem cells to pre-endocardial mesenchyme
3	Atrioventricular and outflow tract endocardium: formation of cushion mesenchyme
4	Epicardium and/or dorsal mesocardium: transformation of epicardial cells as possible precursors for the coronary vessels
5	Neural fold ectoderm: migration of cardiac neural crest gives rise to the aortic-pulmonary septum
6	Coelomic (pulmonary) mesothelium: formation of the spina vestibuli

At Hamburger and Hamilton (H.H.) [1951] stage 4/5 in the chick, the heart-forming fields are composed of loosely aggregated mesodermal cells. These cells progressively condense over the next few hours of incubation to form a true epithelium by H.H. stage 6, the precardiac mesoderm or heart-forming epithelium [Peng et al., 1990; Linask and Lash, 1993]. In vitro studies using primary cultures of the precardiac mesoderm [Linask and Lash, 1993] or a cell line (QCE-6) [Eisenberg and Bader, 1995] derived from the heart fields indicate that the precardiac epithelium is bipotential as it gives origin to both endocardial and myocardial lineages. It requires the second of the epithelial transformations critical to heart development listed in table 1 to completely segregate these two lineages.

Fig. 1. Conceptual representation of three cardiogenic epithelial-mesenchymal transformations. (1) Epithelial cells of the ectoderm (epiblast) transform (gastrulate) into mesoderm at the embryonic midline axis (primitive streak). From the primitive streak, transformed cells migrate to form the mesoderm. (2) Within this mesoderm, the paired heart-forming fields are stratified bipotential epithelial cells (precardiac mesoderm) that will give origin to both endocardial and myocardial lineages. (3) Some cells of the precardiac mesodermal epithelium transform into mesenchymal cells which invade the extracellular matrix beneath the endoderm and express pre-endocardial antigens (QH1, JB3). Persisting, i.e. nontransforming, cells of the precardiac mesoderm ultimately differentiate into myocardial precursor cells. Following segregation of the endocardial precursors, two concentric epithelia (myocardium, endocardium) develop within the fused heart fields. (4) Within specific segments of the primary heart tube, the myocardium secretes an extracellular signal (ES proteins) that induces adjacent QH1+, JB3+ endothelial cells (darkly-shaded) to transform into intracardiac cushion mesenchyme which retains endothelial markers.

1

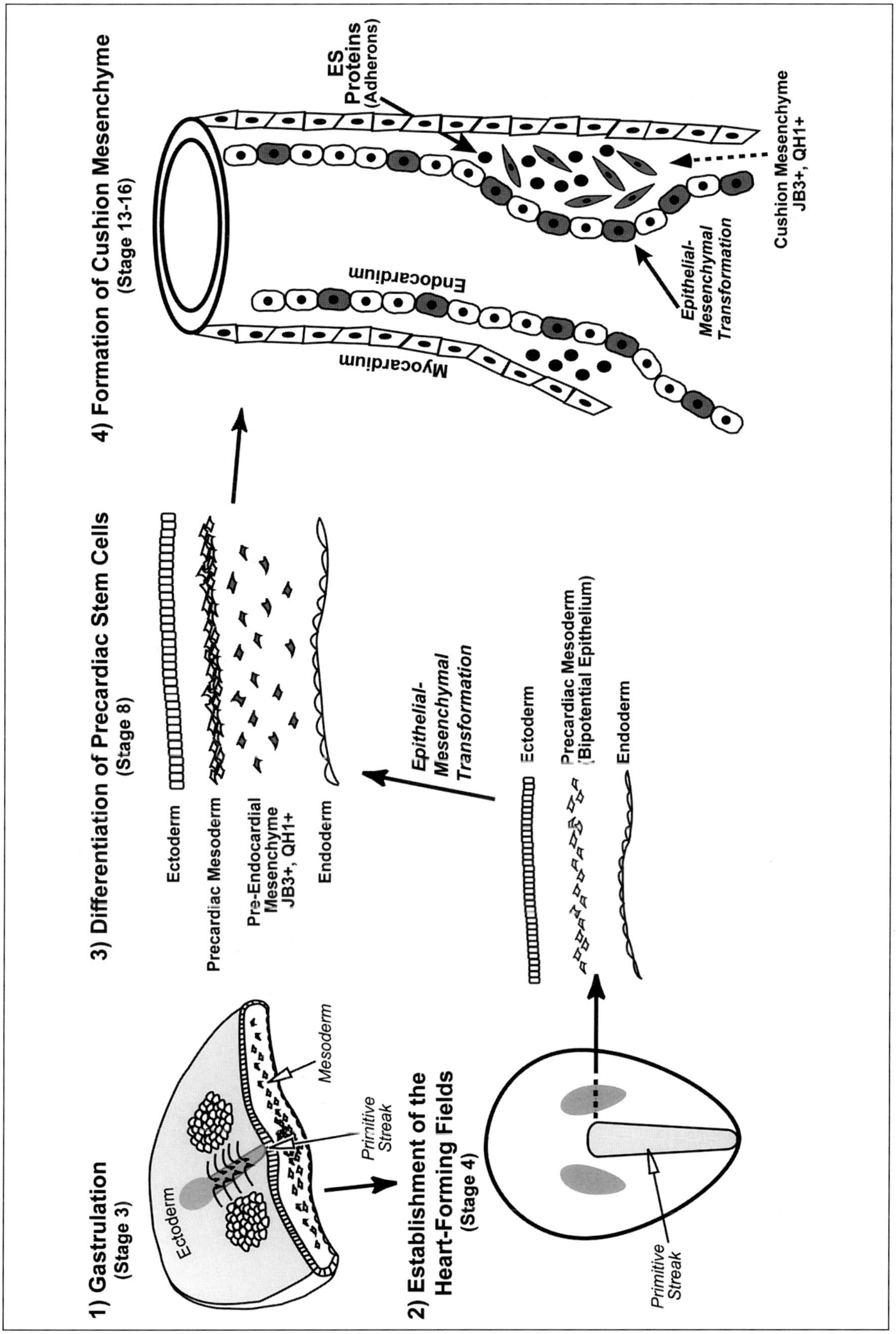
1) Gastrulation
(Stage 3)
Ectoderm
Primitive
Streak
Mesoderm
2) Establishment of the
Heart-Forming Fields
(Stage 4)
Primitive
Streak
3) Differentiation of Precardiac Stem Cells
(Stage 8)
Ectoderm
Precardiac Mesoderm
Pre-Endocardial
Mesenchyme
JB3+, QH1+
Endoderm
Epithelial-
Mesenchymal
Transformation
Ectoderm
Precardiac Mesoderm
(Bipotential Epithelium)
Endoderm
4) Formation of Cushion Mesenchyme
(Stage 13-16)
ES
Proteins
(Adherons)
Endocardium
Myocardium
Epithelial-
Mesenchymal
Transformation
Cushion Mesenchyme
JB3+, QH1+

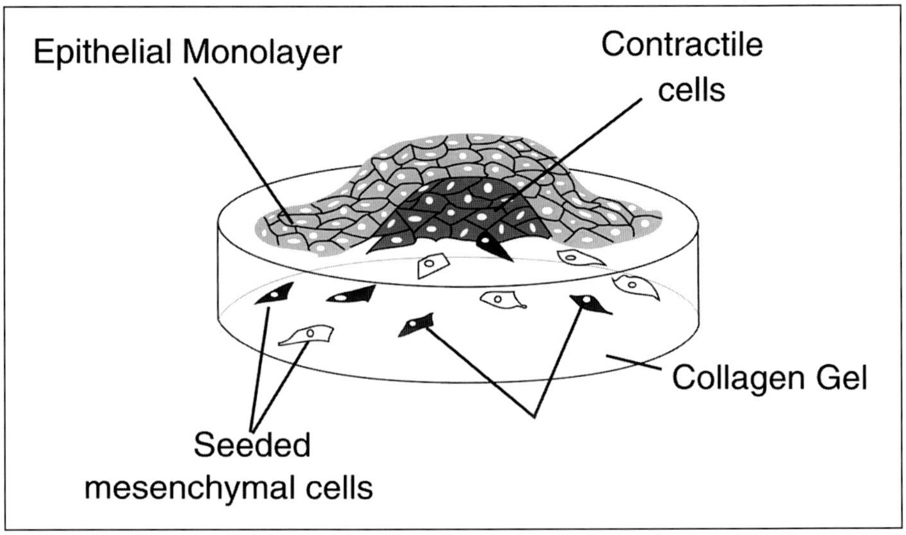

Fig. 2. Typical outcome for culturing precardiac mesoderm (H.H. stage 5/6) on a collagen gel in the presence of endoderm conditioned medium. After 24 h, an epithelial monolayer has formed on the surface of the gel which has seeded (i.e. undergone transformation into) mesenchyme. The latter migrate into the gel lattice and some, but not all, express endothelial markers QH1 and JB3. Contractile cells develop within the original explant. In the absence of endoderm, contractile cells develop but the outgrowth of the epithelial monolayer is truncated, no mesenchyme is seeded, and endothelial antigens are not expressed.

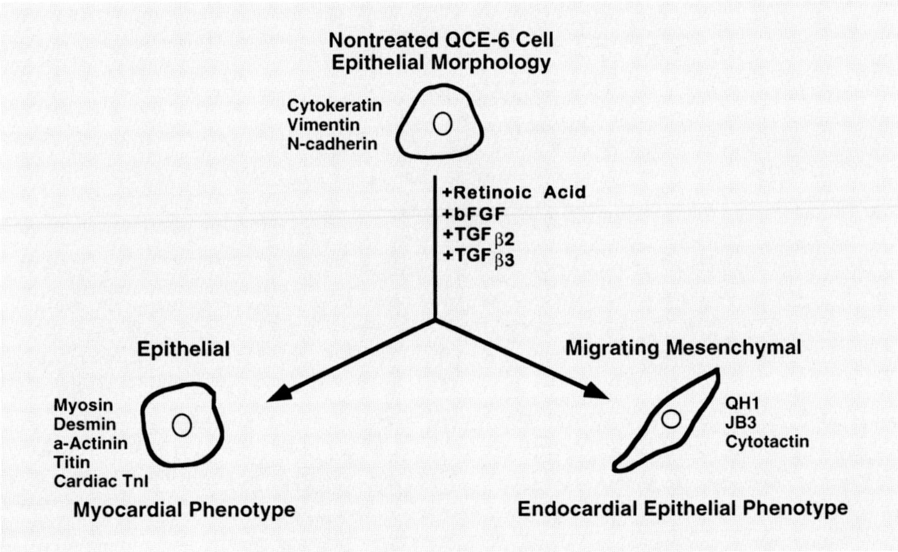

Fig. 3. Procedure for inducing bipotential QCE-6 cells into distinct myocardial and endocardial lineages. As in vivo, myocardial cells retain epithelial-like characteristics, whereas, endocardial precursors are mesenchymal and express endothelial markers, e.g. QH1 and JB3. Retinoic acid and growth factors (FGF, TGFβ) are required to obtain both phenotypes. TnI = Troponin I.

Transformation of Precardiac Mesoderm into Mesenchyme (fig. 1)

Upon splitting of the lateral plate mesoderm to form the coelom (H.H. stage 5/6), the precardiac mesoderm becomes part of the splanchnic mesoderm. Although the splanchnic mesoderm is a major site for the formation of endothelial precursors (angioblasts), the first cells to express the quail endothelial marker – QH1 [Pardanaud et al., 1987] – in the early embryo are mesenchymal cells located in the expansive extracellular matrix separating the precardiac mesoderm and anterior endoderm [Sugi and Markwald, 1996]. We have also found that this population of pre-endothelial mesenchyme expresses the JB3 protein whereas endothelial precursors formed outside the heart forming fields do not express this antigen [Wunsch et al., 1994]. This suggests that endothelial precursor cells may have different developmental fates depending on their origin, an hypothesis consistent with the potential of some endothelial cells to transform 'back' to mesenchyme after having established the inner epithelium (endocardium) of the primary heart tube (table 1).

The development of a 3-dimensional collagen gel culture system has enabled us to study the cellular and molecular events associated with the transformation of different cardiac epithelia into mesenchyme [Bernanke and Markwald, 1982; Markwald et al., 1984]. We have recently used this collagen culture model to verify that a transformation of precardiac mesodermal epithelium is required to establish a population of endocardial precursor mesenchyme

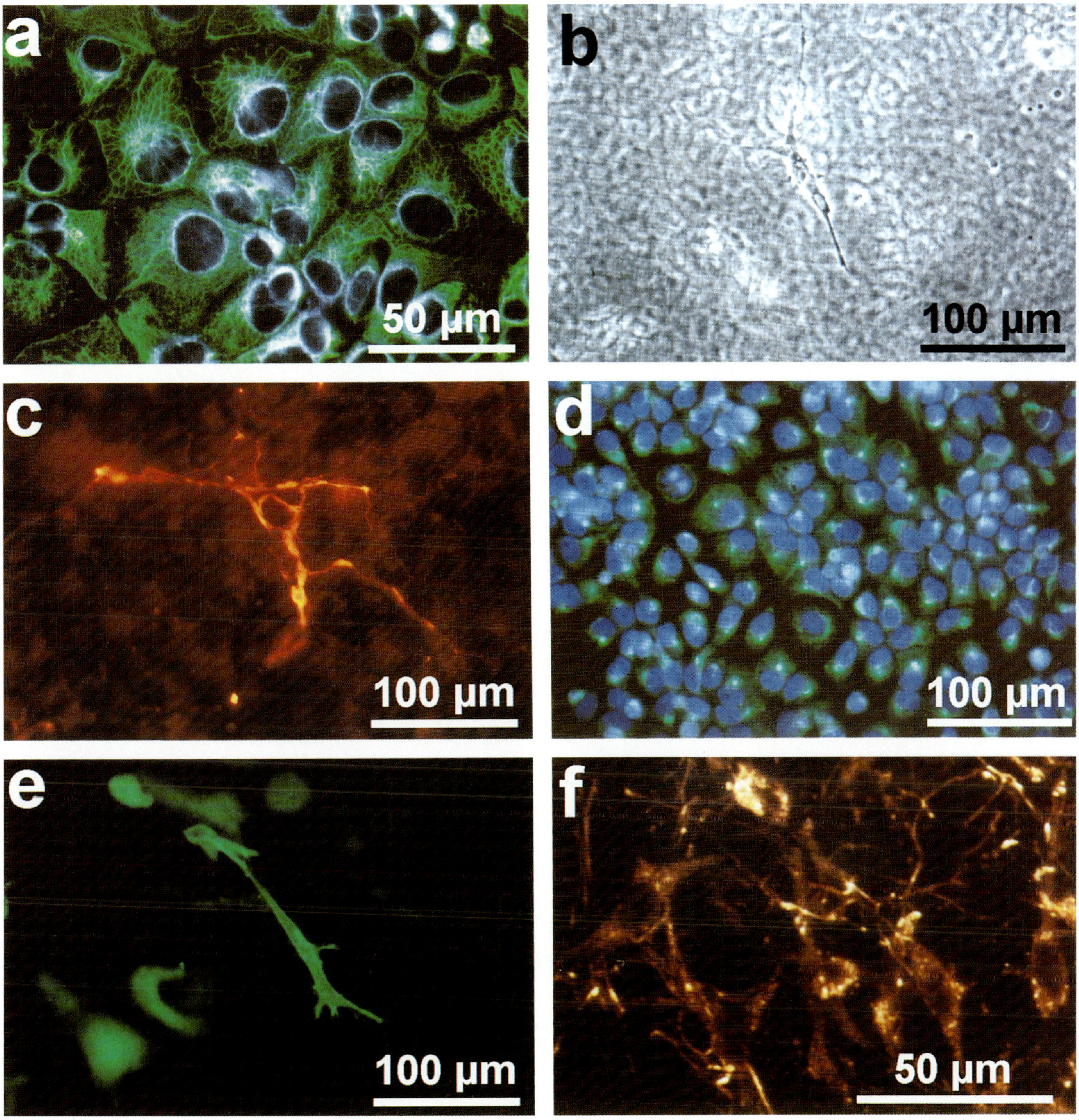

Fig. 4. **a** Noninduced, QCE-6 stem cells stained with antikeratin antibodies; note epithelial morphology and undifferentiated phenotype which is characteristic of primitive mesoderm. **b** QCE-6 cells induced with retinoic acid and growth factors; mesenchymal cells develop which migrate over persisting epithelial cells (myocardial lineage). **c** Same as **b** except stained with anticytotactin antibodies; note only mesenchymal cells express cytotactin. **d** QCE-6 cells induced with only retinoic acid (no growth factors) and stained with anti-ES/130 antibodies. Most cells express ES130 cytoplasmically, particularly in the Golgi, and at their cell surfaces. **e**, **f** Mesenchymal cells derived from the transformation of QCE-6 cells cultured on a collagen gel lattice and induced with TGFβ and basic FGF and stained with QH1 (**e**) or JB3 (**f**) antibodies.

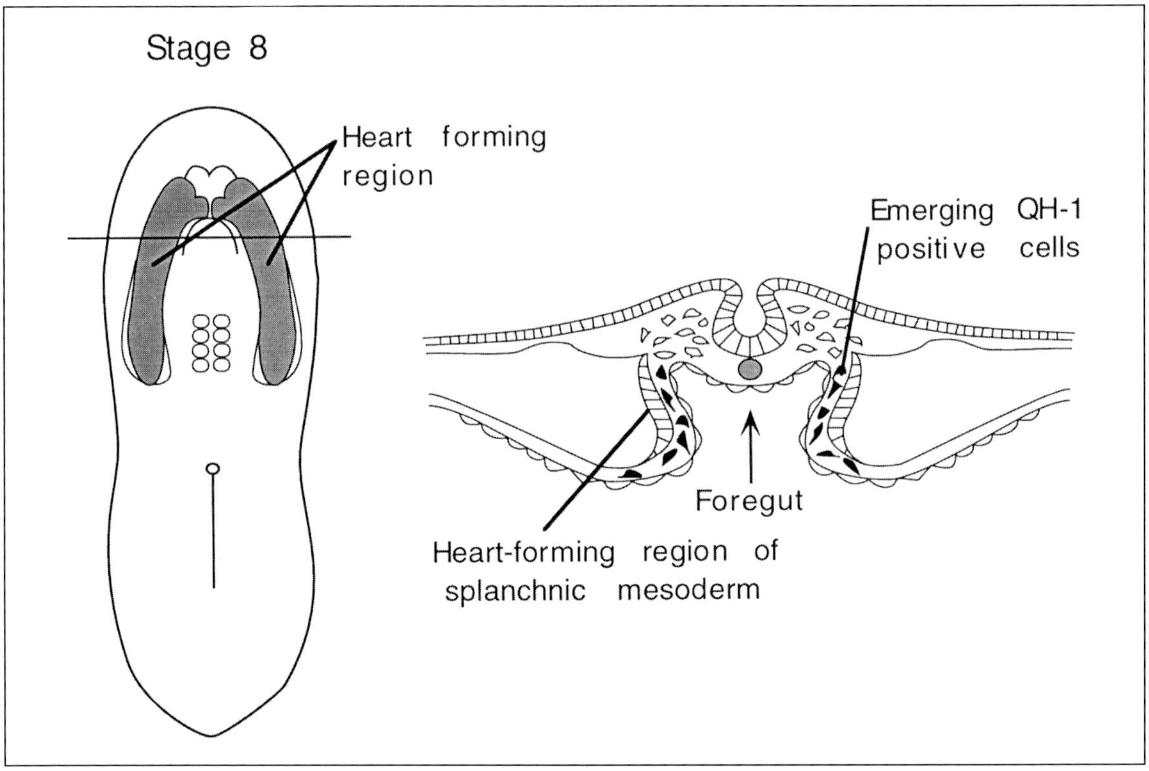

Fig. 5. Left: At stage 8 (four somites), the paired heart fields fuse anteriorly beneath the foregut. Right: A cross-section of a stage 8 embryo at the level shown in the left diagram. Each heart field, prior to fusion, is composed of a heart-forming epithelium (future myocardium) and associated QH1+ (and JB3+), pre-endocardial mesenchyme located within the extracellular matrix associated with the endoderm.

(fig. 2). Specifically, the heart fields from H.H. stage 5/6 embryos are explanted to the top of a hydrated collagen gel lattice in the presence and absence of associated anterior endoderm or endoderm-conditioned medium. In the absence of endoderm, only a few epithelial-like cells grow out from the central primary explant, none of which express the QH1 marker after 4 days in culture. Also, no cells invaded the gel under these conditions [Sugi and Markwald, 1996]. Alpha smooth muscle actin expression and contractile cells were restricted to the cells of the original explant which did not migrate as part of the surface monolayer [Sugi et al., 1993]. Conversely, in the presence of endoderm conditioned medium (fig. 2), the outgrowth of the epithelial monolayer was extensive and, importantly, beneath this monolayer, mesenchymal cells had invaded the gel after 2 days. Many of the newly-formed mesenchymal cells expressed both QH1 and JB3 antigens [Sugi and Markwald, 1996]. These data suggested to us that (i) a segregation of endocardial from myocardially committed cells occurs within the heart forming (precardiac) epithelium; (ii) the segregation of the lineages includes an epithelial transformation into pre-endocardial mesenchymal cells, and (iii) this transformation is dependent upon endoderm. How the endoderm promotes this event is unknown except that the effect is probably through a secreted signal. A number of candidate growth factors are secreted by endoderm that are known to promote endothelial proliferation, migration and differentiation including peptides encoded by the transforming growth factor-β (TGFβ) gene family (e.g. activin, inhibin) and fibroblast growth factor (FGF) [Sugi and Lough, 1995].

The QCE-6 cell line is an important new opportunity to study the transformation of precardiac mesodermal epithelium into pre-endocardial mesenchyme [Eisenberg and Bader, 1995; Eisenberg and Markwald, 1996]. QCE-6 cells are a cloned cell line of cardiac progenitor cells established from stage 4, quail, precardiac mesoderm treated with 20-methylcholanthrene. They exhibit an epithelial phenotype on plastic or collagen gels and express markers characteristic of undifferentiated, precardiac mesoderm (e.g. cytokeratin, vimentin). However, upon addition of retinoic acid and specific growth factors (fig. 3), QCE-6 cells will differentiate into two distinct cardiac cell types, with myocardial

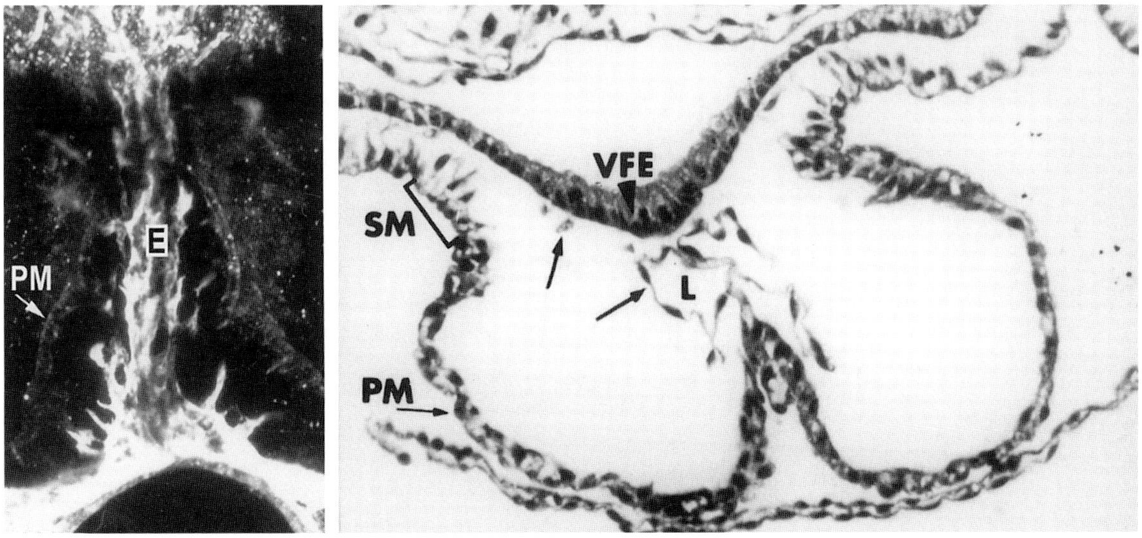

Fig. 6. Left: Stage 9 embryo stained as a whole mount with anti-QH1 antibody, examined by confocal microscopy and analyzed by computer 3-dimensional reconstruction. The primary heart tube is seen extending anteriorly from the QH1+, fused heart fields. Within the heart tube, the developing endocardium (E) is strongly QH1+ and is encased within the unstained mantle of precardiac mesoderm (PM), the future myocardium. Right: A cross-section of similarly staged embryo at a level just rostral to where the two heart fields have fused. Pre-endocardial mesenchymal cells (arrows) integrate to form a single vascular channel or lumen (L). The developing endocardium is separated from the precardiac mesoderm (PM) (presumptive myocardium) by an extracellular matrix (presumptive cardiac jelly) that is continuous with the basement membrane of the ventral foregut endoderm (VFE). The bracketed region of the splanchnic mesoderm (SM) is the junction between precardiac and noncardiogenic mesoderm. ×230.

and endocardial cell properties (fig. 3, 4). When cultured in serum-free defined medium on fibronectin-coated plastic, some, but not all the treated QCE-6 cells transform into mesenchymal cells that express cytotactin and migrate over the persisting epithelial cells (fig. 4a–d). If cultured on a collagen gel, the induced QCE-6 cells form mesenchymal cells that invade the gel lattice and express endothelial markers including QH1 and the JB3 antigen, indicating an endocardial lineage (fig. 4e, f). It may be noteworthy that the growth factors required to induce QCE-6 cells into an endocardial phenotype are also secreted by endoderm. In preliminary experiments, these factors (TGFβ, FGF) will also induce primary cultures of precardiac mesoderm to transform into pre-endocardial mesenchyme [Sugi and Markwald, unpubl. data].

Following the initial formation of pre-endocardial mesenchyme, the paired heart forming primordia – now composed of both epithelium and mesenchyme – fuse at the ventral midline beneath the foregut endoderm (fig. 1, 5). The primary tubular heart develops from the fused heart-forming fields (fig. 6). The pre-endocardial mesenchymal cells interact to form the inner epithelium, the endocardium. Thus, the vasculogenic process leading to the formation of the endocardium can be envisioned as a 'reversal' of the more frequent transformation of epithelium to mesenchyme. The persisting nontransforming cells of the original, precardiac mesoderm differentiate into the myocardial mantle, which retains many epithelial characteristics [Manasek, 1968; Linask, 1992; Byeon et al., 1995]. Cardiac jelly, the extracellular matrix of the primary heart tube, progressively accumulates between the two epithelia (fig. 1, 6, 7). Initially, part of the extracellular matrix associated with the endoderm (fig. 6), cardiac jelly contains 50+ proteins, most of which are secreted by the myocardium after formation of the tubular heart [Krug et al., 1985; Kitten et al., 1987; Little and Rongish, 1995]. For this and other reasons, we have likened the cardiac jelly to an extended basement membrane of the myocardium [Kitten et al., 1987].

Endocardial Endothelial Transformation to Cushion Mesenchyme (fig. 7)

In specific regions along the rostrocaudal (anterior/posterior) axis, the cardiac jelly expands to form the precursor of the endocardial cushions. This expansion of cardiac jelly

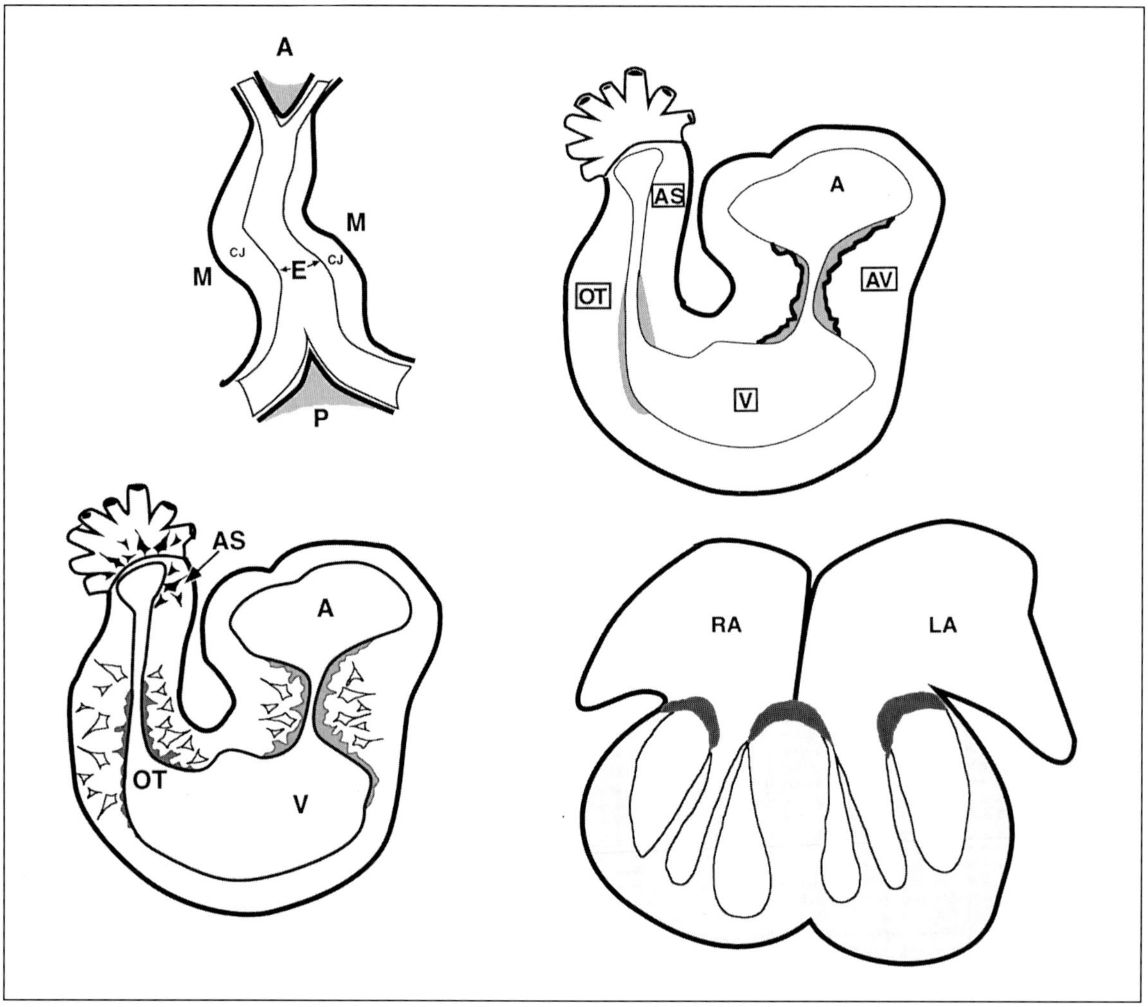

Fig. 7. Diagrammatic representations of the remodeling of the early heart tube into a multichambered organ. Upper left, the primary heart tube is composed of concentric epithelia, myocardium (M) and endocardium (E), separated by cardiac jelly (CJ). Wedges of extracardiac mesenchyme (shaded areas) push into the heart tube at either end; the aortico-pulmonary septum of neural crest origin invaginates into the rostral or anterior (A) pole, whereas spina vestibuli invades at the caudal or posterior (P) pole. Upper right and lower left: the transformation of endocardium to cushion mesenchyme occurs in the atrioventricular (AV) canal and outflow tract (OT), but not the ventricle (V). Neural crest precursors of the aorticopulmonary septum are at the arterial pole within the aortic sac (AS). Lower right: the cushion contributions to the inlet valves (darkly shaded areas) of the four-chambered heart are shown. RA = Right atrium; LA = left atrium.

heralds the only epithelial-mesenchymal transformation to occur entirely within the primary heart tube (fig. 7). The transformation of endocardial endothelium into cushion mesenchyme is restricted to specific segments within the primary tube, the atrioventricular (AV) canal and the conotruncus or outflow tract (the segment connecting the ventricles to the ventral aorta/aortic sac). Within these segments, cushion mesenchyme differentiates into valvuloseptal precursor cells (fig. 8) [Lamers et al., 1995; Wessels et al., 1996]. In simplistic terms, cushion mesenchyme established a blueprint for remodeling the simple heart tube with a common lumen into a four-chambered organ [Van Mierop, 1976; Markwald et al., 1977].

The transformation of endocardium into cushion mesenchyme raises two obvious questions that probably pertain to other epithelial-mesenchyme transitions: (i) why do only some cells within the epithelium transform, and (ii) is the transformation regulated by intrinsic or extrinsic mecha-

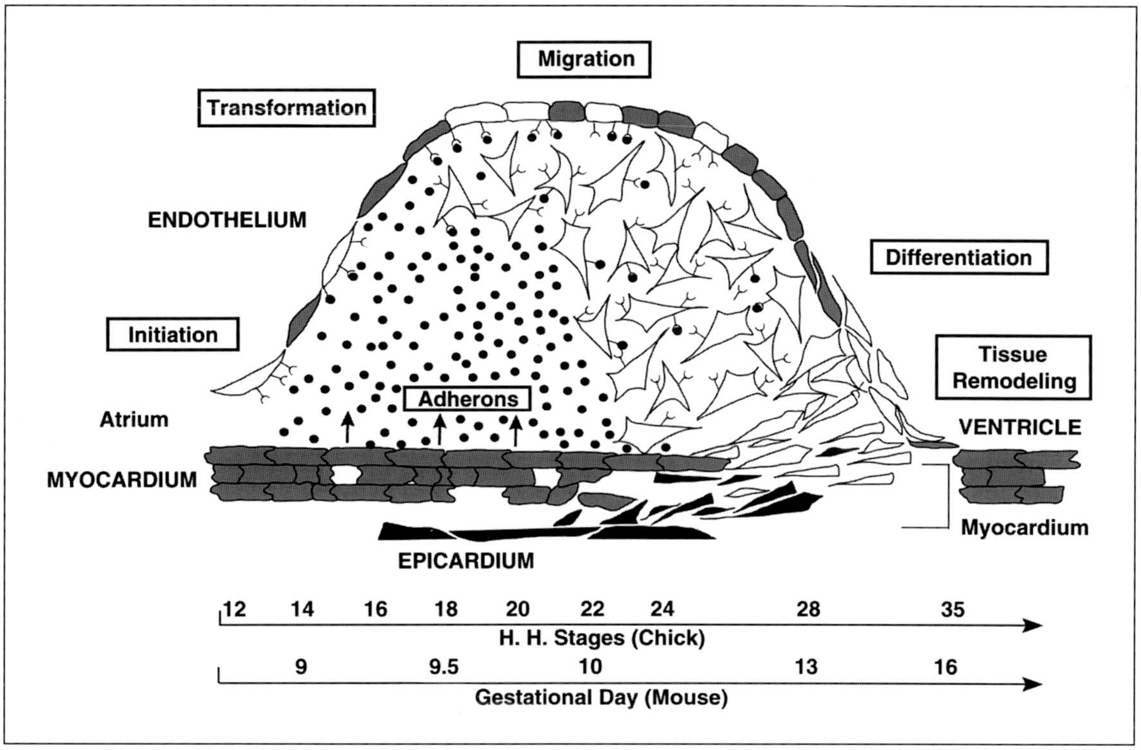

Fig. 8. Conceptual summary of endocardial cushion tissue formation in the AV canal. Three central hypotheses are presented: (i) the endocardium is composed of two subpopulations of endothelial cells, only one of which transforms (clear cells) into cushion mesenchyme; (ii) the AV myocardium induces competent endothelial cells to transform by secreting adherons (complexes of ES antigens and fibronectin) into the cardiac jelly, and (iii) competent endothelial cells have receptors for adheron proteins. A third epithelium, the epicardium, forms at the surface of the heart tube and, like the endocardium, undergoes transformation to mesenchyme. The two populations of mesenchyme directly interact following remodeling of the AV myocardium. The two populations of mesenchyme differentiate into valvular tissues [Wessels et al., 1996]. A chronological chart comparing chick and mouse development is included.

nisms? In figure 8, we present our hypothesis that there are two subsets of endocardial endothelial cells but only one is competent to respond to an extrinsic signal by transforming into mesenchme. To test this hypothesis, we prepared monoclonal antibodies against cell surface antigens isolated from transformed endothelial cells (i.e. the cushion mesenchyme). The antibodies were screened and selected on the basis of their potential to heterogeneously label endocardial endothelial cells. One such antibody – JB3 – recognized only endothelial cells that subsequently transformed to mesenchyme [Wunsch et al., 1994]. Even within the cardiac segments that formed cushion mesenchyme, JB3+ cells were the minority endothelial cell type indicating that only a small percentage (~10%) of endocardial endothelial cells are in a cushion cell lineage. As described above, results obtained by staining the heart fields with JB3 antibody indicated that precursors for JB3+ endocardial endothelial cells most likely were derived from the transformation of heart field epithelium. If true, it would mean that the majority of future endocardial cells are formed external to the heart fields as suggested initially by Coffin and Poole [1991]. Subsequently, the two subsets, JB3+ and JB3– cells would ultimately come together at the ventral midline to form the definitive endocardium (fig. 1). Thereafter, only JB3+ cells would be intrinsically competent to transform to mesenchyme.

The collagen gel culture system [Markwald et al., 1984] has also enabled us to determine that even though some endocardial endothelial cells appear intrinsically competent to transform to mesenchyme, to do so they require an extrinsic signaling mechanism [Krug et al., 1987]. The source of the signal is the myocardium, specifically the myocardial cells of the same segment [Mjaatvedt et al., 1987]. This provides one explanation of why AV canal and outflow tract are the regions of the heart which form the endocardial cushion mesenchyme. The myocardial signal is secreted as part of a multicomponent complex which we called adherons (fig. 8) based on a precedent in the literature [Schubert et al., 1983]. The 'myocardial adheron' can be extracted from the cardiac jelly using EDTA and fractionated from EDTA extracts by high-speed centrifugation. Directly adding pelleted fractions enriched in adherons to cultures of AV or outflow tract endocardial endothelium will induce some AV or outflow tract cells to transform to cushion mesenchyme [Mjaatvedt and Markwald, 1989]. Antibodies made against pelleted adherons – called ES (EDTA soluble) antibodies – block the transformation of endothelium to mesenchyme when added to the culture medium [Mjaatvedt et al., 1991; Nakajima et al., 1994]. When used to immunostain the primary heart tube, anti-ES and antifibronectin antibodies codistribute to a unique form of particulate matrix found only in the cardiac jelly of segments that form cushion mesenchyme [Mjaatvedt et al., 1987, 1991]. Ultrastructurally, the particulate form of matrix appears to be aggregates of adherons, visualized as small 20- to 30-nm particles containing fibronectin and ES antigens [Mjaatvedt and Markwald, 1989]. We have also observed ES antigen-enriched, particulate matrix at other sites of epithelial mesenchymal transformation (e.g. neural crest, somite, precardiac mesoderm) [Brauer and Markwald, 1988; Mjaatvedt et al., 1991]. This suggests that the ES proteins of adherons may have global importance for early mesenchyme formation.

We have used ES antibodies to initiate identification of ES antigens by various molecular approaches. For example, by immunopurification and microsequencing, we have identified one ES antigen as transferrin [Isokawa et al., 1994], a protein that can function as a mesenchymal cell growth factor [Ekblom et al., 1983]. By screening a cDNA expression library prepared from stage 14–16 embryonic heart cells, we have identified another ES antigen as a 130-kD protein which bears no recognizable homology to any known protein. ES130 specific antibodies or antisense oligonucleotides inhibit transformation of endothelium to mesenchyme in the culture assays [Rezaee et al., 1993; Markwald et al., 1995]. ES130 is first expressed in the heart-forming fields prior to their fusion. Therefore, it is not surprising that it is also expressed by QCE-6 cells induced by retinoic acid (fig. 4). Within the primitive heart tube, expression of ES130, both mRNA and protein, is restricted initially to the myocardium. After induction, myocardial expression decreased but responder endothelial cells and newly formed mesenchymal cells both expressed ES130, the latter an unexpected outcome [Markwald et al., 1995].

The upregulated expression of ES130 by induced endothelial/mesenchymal cells makes ES130 a particularly attractive candidate for a homogenetic inducer. 'Homogenetic induction' is the process originally proposed by Spemann [1938] whereby induced responder/target cells become, in turn, signaling cells in either a paracrine or autocrine mode. If true, this hypothesis would indicate that transforming epithelial/mesenchymal cells contribute inductive signals to sustain their own transformation. Consistent with this hypothesis were obervations initially made during the development of the culture system. For example, if myocardial tissue of an AV explant is removed following the initial appearance of transforming endothelial-mesenchymal cells, the transformation process continues, despite the absence of the stimulator tissue [Runyan and Markwald, 1983]. It remains to be determined if the same inductive signal, e.g. ES130, can serve as both a paracrine signal secreted by the stimulator tissue and as an autocrine/paracrine signal secreted by the responder tissue. It also remains unknown as to whether ES130 functions individually or in concert with other proteins within adherons to directly or indirectly regulate endothelial transformation to cushion mesenchyme. Preliminary immunohistochemical findings for ES130 at other sites in the embryo suggest a potential pivotal role for this novel protein in homogenetic, inductive tissue interactions. For example, ES130 is expressed initially in the notochord at H.H. stage 9 and later by two known targets of notochord induction, the ventral midline (floorplate) cells of the developing neural tube and the somites [Krug et al., 1996].

The ability of ES antigens to promote endothelial transformation to mesenchyme evokes questions as to the regulatory mechanism(s) involved. What are the inducible, target cell responses which mediate transformation from an epithelial to mesenchymal phenotype? A number of changes in morphology or gene expression have been observed in transforming endocardial cells following myocardial induction. These include cell hypertrophy, loss of cell:cell adhesions, formation of migratory appendages and polarization of the Golgi apparatus to the future trailing end of the transforming endothelial cell [Krug et al., 1985, 1987]. Diminished N-CAM expression correlates with loss of cell:cell adhesion [Mjaatvedt and Markwald, 1989; Byeon et al., 1995] while the enhanced expression of urokinase [McGuire and Orkin, 1992], cytotactin [Crossin and Hoffman, 1991], sulfated proteoglycans [Funderburg and Markwald, 1986] and fibulin [Spence et al., 1992; Zhang et al., 1992] accompanies migration as a mesenchymal cell. It is unlikely that each of the events requires a specific inductive stimulus. More likely, these and other changes in morphology or gene expression are downstream responses to the induction by ES antigens of one or more key endothelial regulatory genes.

One such candidate gene might be TGFβ peptides [Potts et al., 1992; Nakajima et al., 1994]. The pattern of their distribution is very consistent with a role in mediating epithelial mesenchymal transformations or interactions [Lehnert et al., 1988]. In the primary heart tube, TGFβ are expressed in AV and outflow tract endothelial cells at the time of cushion mesenchyme formation [Akhurst et al., 1990; Nakajima et al., 1994]. In culture assays, endothelial expression of TGFβ-3 is upregulated by ES antigens [Nakajima et al., 1994]. Culture studies using antisense oligonucleotides and antibodies have shown that TGFβ-3 is required for myocardially induced AV endothelium to initiate transformation to mesenchyme [Potts et al., 1990]. The growth factor accumulates primarily at the surfaces of induced endothelial cells and migrating mesenchymal cells suggesting that the secreted growth factor may be bound to cell-surface receptors [Nakajima et al., 1994]. Collectively, these observations suggest that expression of TGFβ is a response by target cells to myocardial ES antigens and that the upregulated growth factor functions in target cells in an autocrine manner to mediate changes in gene expression related to adhesion and migration. Since endothelial ES130 is similarly upregulated by myocardial induction, it is tempting to hypothesize that ES130 protein sustains the expression of endothelial TGFβs which initially were induced by myocardial ES130. To test this hypothesis of homogenetic induction will require the availability of purified ES130 protein, a realistic possibility given the availability of QCE-6 cells (fig. 4d).

Epithelial Transformations Leading to the Formation of Extracardiac Mesenchyme

In table 1, we listed three other epithelial transformations which play key roles in establishing coronary vascular precursors and progenitors of septa that divide the inlet (venous) and outlet (arterial) poles of the primary heart tube (fig. 7). Although listed last, it should be understood that these transformations probably occur concurrently with those of the endocardial endothelium.

Epicardial Mesenchyme

The primitive heart tube eventually acquires a third epithelium, the epicardium. The epicardium forms on the surface of the myocardial epithelium as a migratory outgrowth of the dorsal mesocardium, specifically from the mesothelium of the pro-epicardial organ [Ho and Shimada, 1978; Poelmann et al., 1993]. The latter is a grape-like cluster of mesothelium located between the sinus venosus and the hepatic parenchyma. Initially, the epicardial epithelium is separated from the myocardium by varying amounts of uncharacterized, acellular extracellular matrix. Mesenchymal cells soon invade this space, their derivation likely being from two possible sources, the epicardial epithelium itself or its progenitor, the mesothelium of the pro-endocardial organ [Runyan and Markwald, 1983; Poelmann et al., 1993]. A contribution from the myocardium was disproven several years ago [Manasek, 1969] and recently confirmed by others [Noden et al., 1995]. We have shown in culture assays that the epicardial epithelium from a stage 18 embryo is capable of transforming into mesenchymal cells which, unlike endothelial-derived mesenchyme, formed elongated chains or cords of cells within collagen lattices. Some cords acquired a lumen indicating possible vessel formation [Bolender et al., 1990]. When established from quail explants, some but not all of these cells expressed the QH1 endothelial marker. These findings prompted us to suggest some epicardial mesenchymal cells are in a vascular lineage and contribute to the formation of coronary vessels, although retroviral labeling studies indicate that coronary vascular endothelial cells can originate within the pro-epicardial organ and subsequently migrate onto the surface of the heart beneath the epicardium [Poelmann et al., 1993]. Thus, as with pre-endocardial mesenchyme, epicardial mesenchyme may be derived from more than one source. As shown in figure 8, the intracardiac cushion mesenchyme and the epicardial mesenchyme eventually come into direct contact in the sulcus of the AV canal following remodeling/displacement of the intervening myocardium. Together, they differentiate into the inlet valvular apparatus, the cushions forming the valve leaflets, the epicardial mesenchyme providing the anchoring fibrous support, the 'cardiac skeleton' [Lamers et al., 1995; Wessels et al., 1996]. Thus, the potential developmental fates of epicardial mesenchyme suggest multiple lineages including endothelial cells and stromal fibroblasts.

Virtually nothing is known about the mechanisms regulating epicardial mesenchyme formation except that its proximity to myocardium invites comparison to myocardial induction of endocardial mesenchyme. Preliminary immunohistochemical findings indicate the presence of diffuse (non-adheron-like) ES antigens in the space between epicardium and myocardium [unpubl. data]. As with the other cardiac-related, epithelial-mesenchymal transformations, the JB3 protein is secreted by epicardial mesenchyme [Wunsch et al., 1994]. This protein appears to be a fibrillin-like or fibrillin-associated protein [Gallagher et al., 1993; Wunsch et al., 1994]. The JB3 protein and fibulin are secreted as extracellular fibrils at many sites of epithelial-mesenchyme transformations including not only those linked to heart development, but somites and neural crest as well [Wunsch et al., 1994]. The organization of JB3 and fibulin into fibrils suggests a potential role for these proteins in either mesenchymal cell migration or as a transitional matrix for directly or indirectly promoting mesenchymal cell differentiation [Sugi and Markwald, 1996].

Aortic-Pulmonary Septum and the Spina Vestibuli

The two final transformations of epithelium to mesenchyme listed in table 1 involve epithelia of noncardiac origin, specifically the neural folds and coelomic epithelium. These two epithelia give origin to neural crest [Dubaud et al., 1995] and coelomic mesenchyme, respectively. Each in turn organizes into a distinct wedge of mesenchyme that 'pushes' into the primary heart tube at its arterial (anterior) and venous (posterior) poles, respectively (fig. 7, 8).

Neural crest forms the aortico-pulmonary septum that divides the aortic sac into the roots of the aorta and pulmonary trunk [Kirby, 1993]. This septum is critical to linking the left-ventricular outlet with the fourth aortic arch and the right ventricular outlet with the sixth arch. Kirby and Waldo [1990] have shown that inhibiting the formation of cardiac neural crest results in persistent truncus (common outflow tract) and aortic arch anomalies.

The spina vestibuli is a lesser known structure whose importance has only recently been appreciated [Asami and Koizumi, 1995]. The origin of the spina vestibuli has not been experimentally confirmed; however, the first mesenchymal cells of the future spina vestibuli appear beneath the coelomic mesothelium between the lung buds and developing sinus venosus [Tasaka et al., 1996]. With the incorporation of the sinus venosus into the common atrium, the spina vestibuli enters the posterior end of the heart tube at its midline. There it forms a wedge of mesenchyme which migrates towards the fused AV endocardial cushions 'dragging' a fold of atrial myocardium behind it, the future primary atrial septum. Thus, formation of the spina vestibular mesenchyme directly affects septation of the atrium [Asami and Koizumi, 1995; Tasaka et al., 1996].

It remains to be determined if the transformation of neural ectoderm or coelomic mesothelium occur by processes similar to those described for the precardiac mesoderm, QCE-6 cells and the endocardial endothelium. For example, are similar growth factors or ES antigens involved? The latter have been observed in the basement membrane surrounding sites of neural crest formation [Brauer and Markwald, 1988; Mjaatvedt et al., 1991]. And, in mice trisomic for chromosome 16 ('Down's mouse'), the formation of the AV cushions and the mesenchyme of the spina vestibuli are developmentally delayed and hypoplastic [Hiltgen et al., 1995; Tasaka and Markwald, unpubl. observations]. These findings point to common mechanisms that may regulate epithelial transformations and hold promise for understanding how perturbations in this fundamental developmental process may result in congenital cardiac malformation.

Acknowledgements

This work was made possible by funding provided by a MERIT award (K37-HL33756) and a program project (HL52813) from the National Heart, Lung, Blood Institute.

References

Akhurst, R.J., S.A. Lehnert, A. Faissner, E. Duffie (1990) TGF beta in murine morphogenetic processes: The early embryo and cardiogenesis. Development *108:* 645–656.

Asami, I., K. Koizumi (1995) Development of the atrial septal complex in the human heart: Contribution of the spina vestibuli; in Clark E, R Markwald, A Takao (eds): Developmental Mechanisms of Heart Disease. Armonk, Futura, pp 255–260.

Bernanke, D., R.R. Markwald (1982) Cardiac cushion morphogenetic events in a three-dimensional collagen lattice culture model. Dev Biol *92:* 235–245.

Bolender, D.L., R.R. Markwald (1991) Endothelial formation and transformation in early avian heart development: Induction by proteins organized into adherons; in Feinberg RN, G Sherer, R Auerbach (eds): The Development of the Vascular System. Issues Biomed. Basel, Karger, vol 14, pp 109–124.

Bolender, D.L., M.D. Olson, R.R. Markwald (1990) Coronary vessel vasculogenesis: Embryonic origins of defective heart development. Ann NY Acad Sci *558:* 340–344.

Brauer, P.R., R.R. Markwald (1988) Specific configurations of fibronectin-containing particles correlate with pathways taken by neural crest cells at two axial levels. Anat Rec *222:* 69–82.

Byeon, M.K., Y. Sugi, R.R. Markwald, S. Hoffman (1995) NCAM polypeptides in heart development: Association with Z discs of forms that contain the muscle specific domain. J Cell Biol *128:* 209–221.

Chen, Y., L. Huang, A.F. Russo, M. Solursh (1992) Retinoic acid is enriched in Hensen's node and is developmentally regulated in the early chicken embryo. Proc Natl Acad Sci USA *89:* 10056–10059.

Coffin, J.D., T.J. Poole (1991) Endothelial cell origin and migration in embryonic heart and cranial blood vessel development. Anat Rec *231:* 383–395.

Crossin, K.L., S. Hoffman (1991) Expression of adhesion molecules during the formation and differentiation of the avian endocardial cushion tissue. Dev Biol *145:* 277–286.
Duband, J.L., F. Monier, M. Delannet, D. Newgreen (1995) Epithelium-mesenchyme transition during neural crest development. Acta Anat *154:* 63–78.
Eisenberg, C.A., D. Bader (1995) QCE-6: A clonal cell line with cardiac myogenic and endothelial cell potentials. Dev Biol *167:* 469–481.
Eisenberg, C.A., R.R. Markwald (1996) The quail mesoderm cell line QCE-6: Diversification into blood, endothelial and myocardial cells when co-cultured with avian blastoderm cells. Submitted.
Ekblom, P., I. Thesleff, L. Saxon, R. Timpl (1983) Transferrin as a fetal growth factor: Acquisition of responsiveness related to embryonic induction. Proc Natl Acad Sci USA *80:* 2651–2655.
Funderburg, F.M., R.R. Markwald (1986) Conditioning of native substrates by chondroitin sulfate proteoglycans during cardiac mesenchymal cell migration. J Cell Biol *103:* 2475–2487.
Gallagher, B.C., L.Y. Sakai, C.D. Little (1993) Fibrillin delineates the primary axis of the early embryo. Dev Dyn *196:* 70–78.
Garcia-Martinez, V., G.C. Schoenwolf (1993) Primitive streak origin of the cardiovascular system in avian embryos. Dev Biol *159:* 706–719.
Hamburger, V., H.L. Hamilton (1951) A series of normal stages in the developing chick embryo. J Morphol *88:* 49–92.
Hiltgen, G., L.L. Litke, R.R. Markwald (1995) Morphogenetic alterations during endocardial cushion development in trisomy 16 (Down's) mouse. Pediatr Cardiol *17:* 21–30.
Ho, E., Y. Shimada (1978) Formation of the epicardium studied with the scanning electron microscope. Dev Biol *66:* 579–585.
Icardo, J. (1988) Cardiac morphogenesis and development. Experentia *44:* 909–913.
Isokawa, K., M. Rezaee, A. Wunsch, R.R. Markwald, E.L. Krug (1994) Identification of transferrin as one of multiple EDTA-extractable extracellular proteins involved in early chick heart morphogenesis. J Cell Biochem *54:* 207–218.
Kirby, M.L. (1993) Cellular and molecular contributions of the cardiac neural crest to cardiovascular development. Trends Cardiovasc Med *3:* 18–23.
Kirby, M.L., K.L. Waldo (1990) Role of the neural crest in congenital heart disease. Circulation *82:* 332–340.
Kitten, G.T., R.R. Markwald, D.L. Bolender (1987) Distribution of basement membrane antigens in cryopreserved early embryonic hearts. Anat Rec *217:* 379–390.
Krug, E.L., C.H. Mjaatvedt, R.R. Markwald (1987) Extracellular matrix from embryonic myocardium elicits an early morphogenetic event in cardiac endothelial differentiation. Dev Biol *120:* 348–355.
Krug, E.L., M. Rezaee, K. Isokawa, D.K. Turner, L.L. Litke, A.M. Wunsch, J.L. Bain, D.A. Riley, A.A. Capehart, R.R. Markwald (1995) Transformation of cardiac endothelium into cushion mesenchyme is dependent on ES/130: Temporal, spatial and functional studies in the early chick embryo. Cell Mol Biol Res *41:* 263–277.
Krug, E.L., R.B. Runyan, R.R. Markwald (1985) Protein extracts from early embryonic hearts initiate cardiac endothelial differentiation. Dev Biol *112:* 414–426.
Lamers, W.H., S. Viragh, A. Wessels, A.F.M. Moorman, R.H. Anderson (1995) Formation of the tricuspid valve in the human heart. Circulation *91:* 111–121.
Lehnert, S.A., R.J. Akhurst (1988) Embryonic expression pattern of TGF beta type-1 RNA suggests both paracrine and autocrine mechanisms of action. Development *104:* 263–273.
Linask, K.K. (1992) N-Cadherin localization in early heart development and polar expression of Na+, K+ ATPase and integrin during pericardial coelom formation and epithelialization of the differentiating myocardium. Dev Biol *151:* 213–224.
Linask, K.K., J.W. Lash (1993) Early heart development: Dynamics of endocardial cell sorting suggests a common origin with cardiomyocytes. Dev Dyn *196:* 62–69.
Little, C.D., B.J. Rongish (1995) Extracellular matrix in heart development. Experentia *51:* 873–882.
Manasek, F.J. (1968) Embryonic development of the heart. I. A light and electron microscopic study of myocardial development in the early chick embryo. J Morphol *125:* 329–365.
Manasek, F.J. (1969) Embryonic development of the heart. II. Formation of the epicardium. J Embryol Exp Morphol *22:* 333–348.
Markwald, R.R., T.P. Fitzharris, F.M. Manasek (1977) Structural development of endocardial cushions. Am J Anat *148:* 85–120.
Markwald, R.R., G.T. Kitten, R.B. Runyan, F.M. Funderburg, D.H. Bernanke, P.R. Brauer (1984) Use of three-dimensional collagen gel culture to study cell:matrix interactions in heart development; in Trelstad R (ed): Role of the Extracellular Matrix in Development. Symposium of the Society of Developmental Biology. New York, Academic Press, vol 42, pp 323–350.
Markwald, R.R., M. Rezaee, Y. Nakajima, A. Wunsch, K. Isokawa, L.L. Litke, E.L. Krug (1995) Molecular basis for the segmental pattern of cardiac cushion mesenchyme formation: Role of ES/130 in the embryonic chick heart; in Clark E, R Markwald, A Takao (eds): Developmental Mechanisms of Heart Disease. Armonk, Futura, pp 185–194.
McGuire, P.G., R.W. Orkin (1992) Urokinase activity in the developing avian heart: a spatial and temporal analysis. Dev Dyn *193:* 24–33.
Mjaatvedt, C.H., E.L. Krug, R.R. Markwald (1991) An antiserum (ES1) against a particulate form of extracellular matrix blocks the transformation of cardiac endothelium into mesenchyme in culture. Dev Biol *145:* 219–230.
Mjaatvedt, C.H., R.C. Lepera, R.R. Markwald (1987) Myocardial specificity for initiating endothelial-mesenchymal cell transition in embryonic chick heart correlates with a particulate distribution of fibronectin. Dev Biol *119:* 59–67.
Mjaatvedt, C.H., R.R. Markwald (1989) Induction of an epithelial-mesenchymal transition by an in vivo adheron-like complex. Dev Biol *136:* 118–128.
Nakajima, Y., E.L. Krug, R.R. Markwald (1994) Myocardial regulation of transforming growth factor beta expression by outflow tract endothelium in the early embryonic chick heart. Dev Biol *165:* 615–626.
Noden, D.M., R.E. Poelmann, A.C. Gittenberger-de Groot (1995) Cell origins and tissue boundaries during outflow tract development. Trends Cardiovas Med *5:* 69–75.
Pardanaud, L., C. Altmann, P. Kitos, F. Dieterlen-Lievre, C.A. Buck (1987) Vasculogenesis in the early quail blastodisc as studied with a monoclonal antibody recognizing endothelial cells. Development *100:* 339–349.
Peng, I., J.E. Dennis, E. Rodriguez-Boulan, D.A. Fischman (1990) Polarized release of enveloped viruses in the embryonic chicken heart: Demonstration of epithelial polarity in the presumptive myocardium. Dev Biol *141:* 164–172.
Poelmann, R.E., A.C. Gittenberger-de Groot, M.M.T. Mentink, R. Bokenkamp, B. Hogers (1993) Development of the cardiac coronary vascular endothelium, studied with antiendothelial antibodies, in chicken-quail chimeras. Circulation Res *73:* 559–568.
Potts, J.D., J.M. Dagle, J.A. Walder, D.L. Weeks, R.B. Runyan (1991) Epithelial-mesenchymal transformation of embryonic cardiac endothelial cells is inhibited by a modified antisense oligodeoxynucleotide to transforming growth factor beta-3. Proc Natl Acad Sci USA *88:* 1516–1520.
Potts, J.D., E.B. Vincent, R.B. Runyan, D.L. Weeks (1992) Sense and antisense TGF beta-3 mRNA levels correlate with cardiac valve induction. Dev Dyn *193:* 340–345.
Rezaee, M., K. Isokawa, N. Halligan, R.R. Markwald, E.L. Krug (1993) Identification of an extracellular 130 kDa protein involved in early cardiac morphogenesis. J Biol Chem *268:* 14404–14411.
Runyan, R.B., R.R. Markwald (1983) Invasion of mesenchyme into three-dimensional collagen gels: A regional and temporal analysis of interaction in embryonic heart tissue. Dev Biol *95:* 108–114.
Schubert, D., M. LaCorbiere, F.G. Klier, C. Birdwell (1983) The structure and function of myoblast adherons. Cold Spring Harbor Symp Quant Biol *48:* 539–549.
Spemann, H. (1938) Embryonic Development and Induction. New Haven, Yale University Press.
Spence, S.G., W.S. Argraves, L. Walters, J.E. Hungerford, C.D. Little (1992) Fibulin is localized at sites of epithelial-mesenchymal transitions in the early embryo. Dev Biol *151:* 73–84.

Sugi, Y., J. Lough (1995) Activin-A and FGF-2 mimic the inductive effects of anterior endoderm on terminal cardiac myogenesis in vitro. Dev Biol, in press.

Sugi, Y., R.R. Markwald (1996) Formation and early morphogenesis of endocardial endothelial precursor cells and the role of endoderm. Dev Biol *175:* 66–83.

Sugi, Y., J. Sasse, J. Lough (1993) Inhibition of precardiac mesoderm cell proliferation by antisense oligonucleotide complementary to fibroblast growth factor-2 (FGF-2). Dev Biol *157:* 28–37.

Tasaka, H., E.L. Krug, R.R. Markwald (1995) Origin of the orifice of the pulmonary vein in the mouse; in Clark EB, RR Markwald, A Takao (eds): Developmental Mechanisms of Heart Disease. Armonk, Futura, pp 347–350.

Tasaka, H., E.L. Krug, R.R. Markwald (1996) Origin of the pulmonary venous orifice in the mouse and its relationship to the morphogenesis of the sinus venosus, extracardiac mesenchyme (spina vestibuli) and atrium. Anat Rec, in press.

Van Mierop, L.H.S. (1976) Morphological development of the heart; in: Handbook of Physiology – The Cardiovascular System, chap 1. Bethesda, American Physiological Society, pp 1–28.

Wessels, A., M.W.M. Markman, J.L.M. Vermeulen, R.H. Anderson, S.Z. Viragh, A.F.M. Moorman, W.H. Lamers (1996) The development of the atrioventricular junction in the human heart: an immunohistochemical study. Circ Res *78:* 110–117.

Wunsch, A.M., C.D. Little, R.R. Markwald (1994) Cardiac endothelial heterogeneity defines vascular development as demonstrated by the diverse expression of JB3, an antigen of the endocardial cushion tissue. Dev Biol *165:* 585–601.

Zhang, H.-Y., M.-L. Chu, P. Te-Cheng, T. Sasaki, R. Timpl, R. Ekblom (1995) Extracellular matrix protein fibulin-2 is expressed in the embryonic endocardial cushion tissue and is a prominent component of valves in adult heart. Dev Biol *167:* 18–26.

Acta Anatomica

Acta Anat 1996;156:187–201

J.A. Davies

Department of Anatomy,
Edinburgh University Medical School,
Edinburgh, UK

Mesenchyme to Epithelium Transition during Development of the Mammalian Kidney Tubule

Key Words
Epitheliogenesis
Kidney
Nephron
Differentiation
Mesenchyme-epithelium transition

Abstract
Kidney epithelia have two separate origins. Collecting ducts develop in the manner of most glandular organs, by growth and branching of a bud derived from a pre-existing epithelium. Excretory tubules develop by a direct mesenchyme to epithelium transition (MET), which is induced by the tips of the developing collecting duct system as it invades a specialised area of mesenchymal cells. The process by which these metanephrogenic mesenchyme cells achieve MET can be divided into several stages; induction, acquisition of stem cell character, fate determination, condensation, epitheliogenesis, polarisation and maturation. Progress through these stages is regulated by 'checkpoints' at which permission to proceed requires specific signals. The stages of development are characterised by the expression of new combinations of genes that code for transcription factors (Hox genes, Pax genes, zinc finger proteins), signalling effectors (growth factors, Wnts, receptor tyrosine kinases) and morphoregulatory molecules (CAMs, cadherins, extracellular matrix ligands). This review summarises current knowledge about the molecular interactions that control MET in the kidney, and also about how their failure might result in Wilms' tumour, one of the most common cancers of childhood.

Introduction

The morphogenesis of most epithelial organs is fairly simple. Typically, a pre-existing epithelial sheet or tube gives rise to the new organ's primary duct, which undergoes successive branching events to generate a tree-like structure. Epithelial growth and arborisation is normally supported by surrounding mesenchyme, which provides mechanical stability and morphogenetic signals but does not contribute cells to the growing epithelium. This simple theme, with minor variations, accounts for the anatomy of organs as diverse as salivary gland, mammary gland, lung and pancreas. In some cases, however, mesenchyme plays an even greater role, and provides the cells from which new epithelium will be built.

The mesenchyme-epithelium transition (MET) necessary for this process is a rare event in development, being seen in only a few tissues (e.g. somite, kidney, testis). Often the tissue so formed is highly unusual and lacks features associated with the epithelial phenotype. For example, somitic epithelium is transitory and expresses adhesion molecules (e.g. N-cadherin) more usually associated with neural cells [Duband et al., 1987] rather than those common in permanent epithelia (e.g. it lacks desmosomal cadherins [Gar-

KARGER

E-Mail karger@karger.ch
Fax + 41 61 306 12 34
http://www.karger.ch

0001–5180/96/1563–0187$12.00/0

Dr. Jamie A. Davies
Department of Anatomy
The University of Edinburgh Medical School
Teviot Place
Edinburgh EH8 9AG (UK)

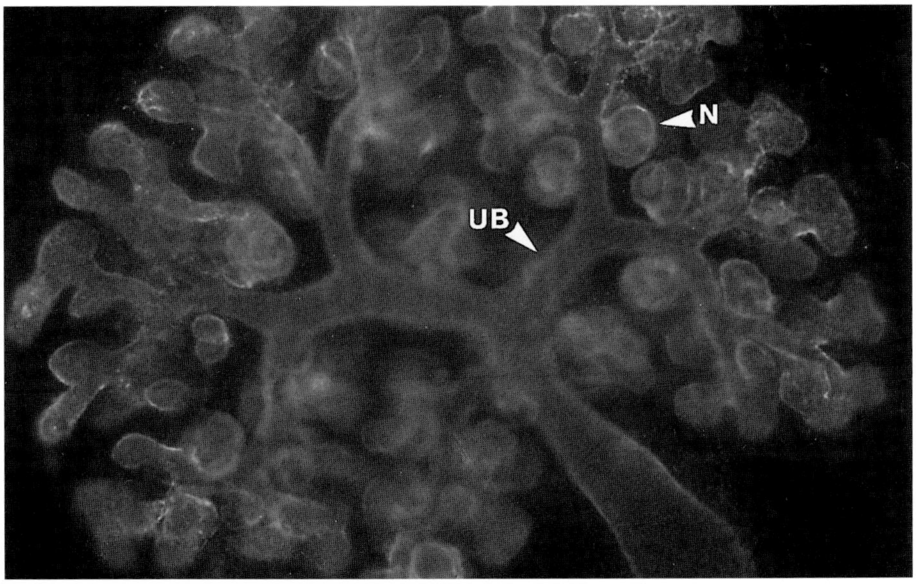

Fig. 1. View of a kidney rudiment grown in organ culture. The ureteric bud (UB) has branched several times, and has induced the formation of nephrons (N) by MET. Anti-laminin stain.

rod, pers. commun.]). Seminiferous epithelium, which is a composite of cells derived from a pre-existing epithelium and those obtained by MET, is organised to 'secrete' entire cells (spermatozoa) into its lumen. The MET of kidney development is particularly remarkable for producing a classical simple transporting epithelium that persists into adult life. This is one of the reasons that it has attracted so much attention in recent decades. The other reason is medical; 10% of us are born with some form of kidney abnormality [Kissane, 1983], and while most are relatively harmless, a few (e.g. renal agenesis, polycystic disease, nephroblastoma) are life-threatening.

In the following, I shall provide a brief overview of kidney development to place its MET in context, and will then describe what we know about epithelial differentiation, primarily in murine kidneys, in detail. It will be obvious that much remains to be done, but it should also be apparent that the field has reached an exciting stage which promises much greater understanding of the molecular mechanisms that control this intriguing aspect of developmental biology.

An Overview of Kidney Development

Kidney development encompasses both 'classical' epitheliogenesis, in which a pre-existing epithelium grows and branches to form a tree, and epitheliogenesis by MET. It begins when a simple epithelial tube called the ureteric bud invades a special area of intermediate mesoderm called the metanephrogenic mesenchyme. Interactions between the two tissues drive the arborisation of the ureteric bud and induce the mesenchyme to undergo a series of differentiation steps that lead ultimately to the formation of epithelial nephrons, around 2,000,000 in man [Dunhill and Haley, 1972].

Kidney rudiments, even those so young that the ureteric bud has not begun to branch, can be identified and removed intact by microdissection. They adapt well to organ culture and follow an almost normal developmental programme (fig. 1), differing from their in vivo counterparts mainly by failing to enlarge at the same rate. It is possible to separate the ureteric bud and metanephrogenic mesenchyme before their reciprocal interactions have taken place, and to place each tissue in organ culture either alone or in contact with other tissues. Such experiments have shown that the ureteric bud cannot develop without contact with metanephrogenic mesenchyme (other mesenchymes won't do), but that metanephrogenic mesenchyme can be induced to undergo MET by a variety of inducing tissues including ureteric bud and embryonic spinal cord [Grobstein, 1953, 1955].

The ability of metanephrogenic mesenchymes to differentiate in culture has provided an excellent system for study of the MET. By controlling the time at which an inducing tissue is placed in contact with the mesenchyme, a researcher can determine accurately the order of consequent events, while development in vitro greatly facilitates manipulation of the system with antibodies, growth factors, etc. Experiments such as these, coupled with observations made on kidneys growing in vivo, have provided much useful information on the various steps involved in the MET.

Each stage of the process so far identified will be discussed in detail below; the essential features of the entire MET are summarised in figure 2.

Starting Point: The Uninduced Metanephrogenic Mesenchyme

The uninduced metanephrogenic mesenchyme appears as a vaguely elliptical group of a few hundred cells at the caudal end of the intermediate mesoderm, its precise boundary being morphologically ill-defined. It gives rise, on induction, to both excretory epithelium (via MET) and renal stroma [Aufderheide et al., 1987]. Does this reflect a choice of fates amongst initially equivalent cells, or are two cell types, one stromogenic and one epitheliogenic, present in the uninduced nephrogenic mesenchyme?

Cell marking experiments have provided some evidence for possible heterogeneity, though its functional significance is unknown. Studies of the migration routes of labelled neural crest cells have revealed that a few find their way into the metanephrogenic mesenchyme area [Weston, 1963; LeDouarin, 1969; Le Douarin and Teillet, 1974]. They probably give rise to the scattered neurons that Sariola et al. [1988] observed in the mesenchymal areas of developing kidneys, rather than to either the epithelial or stromal tissues. Studies of diI-labelled ureteric bud epithelia, reported in a recent review [Herzlinger et al., 1993], appear to demonstrate detachment and integration of bud-derived cells into the mesenchyme followed by their subsequent recruitment into the epitheliogenic pathway. Viewed in isolation, this result might suggest that the ureteric bud, rather than the mesenchyme itself, is the source of epitheliogenic cells. However, other data refute this; nephrogenic mesenchymes from mutant (c-ret −/−) mice that have no ureteric buds can still be induced in culture to form nephrons by MET [Schuchardt et al., 1994; Pachnis, pers. commun.]. Contribution of ureteric bud cells is therefore unnecessary, and its significance remains a mystery for the moment.

Creation of the metanephrogenic mesenchyme requires the transcription factor WT-1, a highly conserved zinc finger protein that has homologues even in plants [Gessler et al., 1990; Rivera-Madrid et al., 1993]. The requirement for WT-1 is shown most dramatically by WT1 −/− mice, in which cells at the site usually occupied by metanephrogenic mesenchyme undergo apoptosis. The mesenchyme is incapable of being induced even if placed manually in direct contact with an inducing tissue [Kreidberg et al., 1993]. Expression of WT1 is maintained at a low level in uninduced cells [Mundlos et al., 1993], but is modulated by later

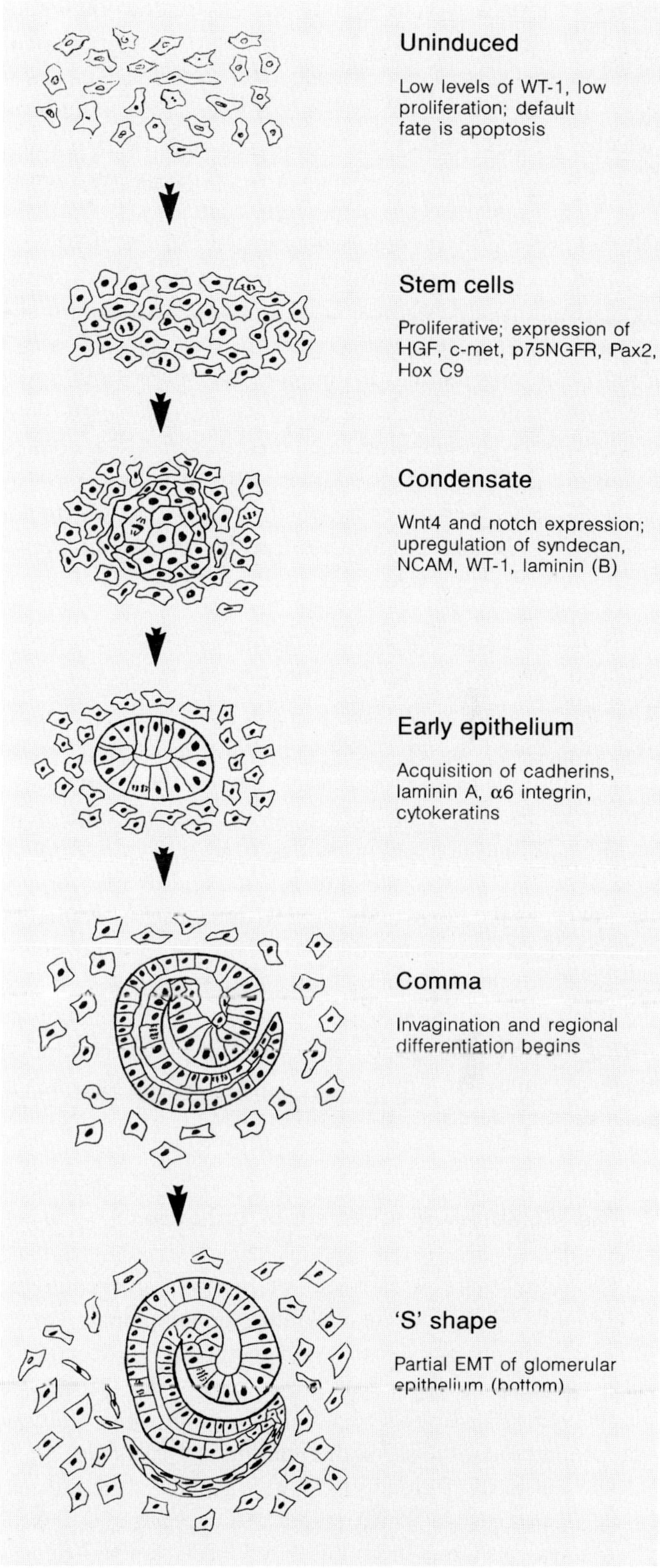

Fig. 2. Stages of MET described in the text.

events of kidney development (see below). No other transcription factors have yet been shown to be essential for creation of the metanephrogenic mesenchyme, but messages for the homeobox-containing gene LFB3 have been detected in these cells [Orr-Otollo et al., 1993]; their function is unknown. A combination of transcription factors presumably specifies the nephrogenic capacity of the cells, which is shared by no other tissue tested [Saxén, 1970].

Uninduced metanephrogenic mesenchyme is short-lived. In normal embryos, it is quickly invaded by the ureteric bud and is induced to enter the next stage of its differentiation programme. If this induction is prevented, the mesenchymal cells undergo programmed cell death. This can be observed when uninduced mesenchymes are isolated in culture [Koseki et al., 1992] and also in vivo: c-ret –/– or *short-tail* (Sd) embryos, in which ureteric bud fails to develop properly and cannot therefore provide an inductive signal, show regression of the metanephrogenic mesenchyme [Schuchardt et al., 1994].

The stark choice of fates available to metanephrogenic mesenchyme cells, to be induced or to die, is incompatible with the pervasive statement that the cortical shell of mesenchyme found in older fetal kidneys, which gives rise to later forming nephrons, consists of naive nephrogenic mesenchyme. To have built such a shell, the progenitor cells would have had not only to survive but to have multiplied extensively in order to maintain and expand their own population as well as providing the raw material for epitheliogenesis. This is not what they do when left to themselves; the arrival of the ureteric bud must therefore have signalled the nephrogenic mesenchyme to acquire a new phenotype – that of the nephrogenic stem cell. This signal is the first stage of 'induction'; more stages are required before the cells become determined to epithelialise.

The Molecular Mechanism of Induction

The signal that passes from ureteric bud to mesenchyme is arguably *the* key event in kidney development, yet its molecular nature has still to be determined. Four decades of study, using a range of techniques from tissue culture to biochemistry, have succeeded only in establishing the general properties of the signalling factors involved.

The earliest information came from tissue recombination experiments, in which uninduced nephrogenic mesenchyme was cultured in contact with other embryonic tissues. Many but by no means all of these were found to be capable of inducing nephrogenic MET, but the surrogate inducers (ureteric bud, submandibular gland epithelium, spinal cord, brain, salivary gland mesenchyme and developing bone) [Grobstein, 1955; Lombard and Grobstein, 1969; Unsworth and Grobstein, 1970; Saxén, 1987] have nothing obviously in common. Attempts to substitute extracts of inducing tissues, or even freshly killed inducing tissue, for the a living inducer have so far failed [Auerbach and Grobstein, 1958; Saxén, 1987], suggesting that the molecule(s) involved might be very short-lived or might act only when presented to the responding cell in a particular way, for example at a very high local concentration.

Induction appears to require cell-cell contact, and can be prevented by separation of inducing and responding tissue by a filter with pore sizes too small to allow penetration of cell processes [Saxén et al., 1976; Saxén, 1980]. However, sites of contact between inducing and responding cells give no clues about the mechanism of their communication; electron microscopy reveals no specializations and little or no extracellular matrix – the membranes simply touch [Wartiovaara et al., 1974; Lehtonen, 1976]. In vitro induction experiments have demonstrated transfer of radiolabelled amino acids and/or peptides from the inducing to the responding tissue, but while this correlates with induction no causal connection has been proven [Koch and Grobstein, 1963].

Some attempts have been made to guess at the identity of the inducing molecule, either on the basis of receptors known to be present, or the known effects of a candidate signalling molecule on other embryonic systems. Lessons learned from the search for inducers in other systems, such as the amphibian embryo, stress the need for careful definition of criteria that a genuine in vivo inducer must meet [Slack, 1993]. These can be adapted for the case of nephron development: the molecule in question must:

(1) Be present in the natural inducing tissue (ureteric bud tips) as long as induction takes place.

(2) Be capable of inducing nephrogenesis by itself, in *completely* uninduced metanephrogenic mesenchyme.

(3) Be essential for induction (i.e. inhibition of the factor should block induction).

No molecule studied to date meets these three criteria. Most (activin, EGF, FGF, HGF, IGFI, IGFII, IL6, LIF, NGF, PDGF, retinoic acid, TGFα, TGFβ) fail to induce MET [Weller et al., 1991; Bard and Ross, 1991; Gilbert, pers. commun.; Sariola, pers. commun.; Davies, unpubl.]. There are preliminary reports (e.g. *Science* news section, vol 266, p 565) of one molecule, Wnt1, that might have inducing activity. Unfortunately, Wnt1 is not expressed by the natural inducer in vivo [Wilkinson et al., 1987], and its transgenic deletion results in defects in the CNS rather than the excretory system [Bradley et al., 1990].

Whatever mediates the extracellular part of the signalling pathway, news of its arrival must be relayed to the recipient's transcription-regulating machinery by a second messenger system. This line of thought led Davies and Garrod [1995] to attempt induction by direct pharmacological activation of the appropriate second messenger system. Isolated mesenchymes were treated with a variety of second messenger modulators (phorbol esters, staurosporine, dibutyryl cyclic AMP, dibutyryl cyclic GMP, lithium chloride, L-690-488, A23187). One, the lithium ion, initiates early stages of nephron differentiation, including MET. Unfortunately, lithium ions affect many aspects of cell metabolism [reviewed by Birch, 1991], and attempts to identify precisely which are active in kidney development have so far failed [Davies and Garrod, 1995]. All is not gloom, however – lithium-mediated induction has at last provided a means to induce differentiation in isolated mouse metanephrogenic mesenchyme without the need for addition of heterologous inducing tissues. This greatly simplifies the interpretation of experiments in which reagents are added to the system (one no longer has to worry whether a result might be due to effects on the inducing rather than responding tissue), and also permits sensitive analytical techniques such as PCR to be applied without risk of false positives from the heterologous cells.

The Stem Cell Population

Following contact with an inducing tissue, the nephrogenic mesenchyme is rescued from its suicidal tendency; the number of apoptotic cells falls, the number undergoing DNA synthesis rises, and the population size begins to increase [Saxén et al., 1983]. The cells also become more closely packed than those in the uninduced metanephrogenic mesenchyme. The rise in DNA synthesis happens within 8 h of in vitro induction [Davies and Garrod, 1995, unpubl. data], but cells are not committed to further differentiation at this stage – if the inducing stimulus (e.g. Li^+) is removed, no tubules are formed. Acquisition of the stem cell phenotype is accompanied by the activation of new genes which have the potential to regulate cell behaviour.

Nephrogenic stem cells express a number of signalling molecules and receptors not expressed by their uninduced predecessors. Some appear to act in an autocrine loop while others have the potential to do so (their role has yet to be tested). The action of these factors is presumably connected with population control in response to conditions in the stem population itself and in the rest of the embryo.

One likely autocrine loop is mediated by hepatocyte growth factor (HGF), a large growth factor that has a variety of mitogenic, motogenic and morphogenetic effects on different cell types [for review see Gherardi et al., 1993; see also Théry and Stern, this vol.]. It is synthesised by nephrogenic stem cells, and may also be made by the uninduced mesenchyme [Woolf et al., 1994]. Its receptor, c-*met*, is also expressed by the stem cells [Karp et al., 1994; Woolf et al., 1994]; and interference with the autocrine loop using specific antibodies results in markedly increased apoptosis in what should be the stem cell population, and prevention of subsequent MET. It is not yet clear whether the effect on the MET is simply a consequence of the failure of population expansion in the stem cells, or reflects a direct requirement for HGF in epithelial differentiation. Hints of a role for HGF in this type of differentiation have been obtained from experiments with non-renal immortal cell lines; if apparently mesenchymal 3T3 cells are transfected with c-met, to create an HGF-mediated autocrine loop, they acquire a more epithelial phenotype [Tsarfaty et al., 1994]. The HGF produced by nephrogenic stem cells has paracrine functions too; interference with its action markedly inhibits development of growth of the ureteric bud tree [Santos et al., 1994; Woolf et al., 1994; Davies et al., 1995].

Midkine may mediate another autocrine signal loop in the stem cells, although evidence for this is as yet only circumstantial. Production of the cytokine has been demonstrated in normal organs [Muramatsu et al., 1993], and a role in population control has been demonstrated in Wilms tumour, a neoplasm of nephrogenic stem cells (see below); inhibition of autocrine midkine action in cultured Wilms tumour cells, using specific antibodies, inhibits proliferation [Muramatsu et al., 1993; Tsutsui et al., 1993]. The cell type G401, which used to be thought to be derived from Wilms tumour but is more likely to be rhabdoid [Garvin et al., 1993], shows a similar autocrine requirement for a small heat-labile protein called nephroblast growth factor (NB-GF) [Burrow and Wilson, 1993]. Too little has been published about NB-GF to comment on its expression in normal kidneys, or its possible relationship to MK.

Neurotrophins may also be important to stem cell regulation, though the evidence here is confusing. Stem cells express the low affinity neurotrophin receptor p75-NGFR [Sariola et al., 1991; Durbeej et al., 1993], and antisense inhibition experiments have shown that p75-NGFR is necessary for kidney development [Sariola et al., 1991; though Durbeej et al., 1993, present an alternative view]. However, transgenic knockout of p75-NGFR results in no apparent defect in kidney development [Lee et al., 1992], showing that neurotrophins are not essential to the process.

Other regulatory signals are provided by the insulin-like growth factor family. Stem cells produce both IGFII and its binding protein IGFBP2 [Matsell, 1993], while epithelia express the receptor for IGFI [Liu et al., 1993; Kanwar et al., 1994]. Antisense inhibition IGFI receptor blocks both ureteric bud development and (consequently?) MET [Liu et al., 1993].

As all of this intercellular signalling takes place, cells change their expression of transcription factors, acquiring the homeobox gene Hox C9 and the paired box gene Pax2 [Erselius et al., 1990; Dressler et al., 1992; Rothenpieler et al., 1993]. The role of Hox C9 has not yet been tested in this system, but Pax2 has been shown to be essential for nephron development. Antisense knockout of the gene prevents MET [Rothenpieler et al., 1993].

Pax2 is completely absent from uninduced cells [Phelps and Dressler, 1993] but present very soon after induction has taken place either in vivo or when lithium ions are used [Rothenpieler et al., 1993; Davies and Garrod, unpubl. data]. The very early expression of Pax2 makes the gene a useful marker for the induced state. This is important because the value of a number of otherwise excellent studies of gene expression has been reduced by uncertainty over whether cells are completely uninduced or are already stem cells. The use of Pax2 as a marker would help to determine which of the following regulatory factors are expressed by uninduced mesenchyme as well as stem cells, and which are expressed in stem cells alone; Hox A3, A4, A5, A6, B5, C6 and C8, and Mox 1, TCF1, LEF1, N-myc, L-myc and p53 [Gaunt, 1987, 1988; Holland and Hogan, 1988; Sharpe et al., 1988; Mugrauer et al., 1991; Schmid et al., 1991; Cardia et al., 1992; Wall et al., 1992; Oesterwegel et al., 1993]. The functions of most of these remain untested. An exception is N-myc, transgenic ablation of which results in a variable failure of kidney development [Stanton et al., 1992].

Commitment to Epithelial Differentiation

If uninduced metanephrogenic mesenchyme is exposed to an inducing stimulus (living tissue or lithium ions) for only a few hours, the mesenchyme acquires stem cell characteristics (e.g. altered DNA synthesis) but does not differentiate further [Saxén and Lehtonen, 1978; Saxén et al., 1983; Davies and Garrod, 1995, unpubl. data]. Only longer treatments (> 12 h for Li^+-mediated induction) result in the initiation of MET, suggesting that a second signalling event (still covered by the traditional term 'induction') might be necessary before the stem cells can proceed [Davies and Garrod, 1995]. This may have exactly the same molecular basis as the first signal (like the first signal, it can be stimulated by lithium treatment), but is required at a different time. Its likely function in vivo is to ensure that only those cells very near the growing ureteric bud quit the stem cell population and differentiate, the rest of the cells being left to carry on cycling.

Requirements for such signals can create potential 'checkpoints', at which cells must await permission to proceed to the next stage of differentiation. Experiments in organ culture have revealed several other stages when particular interactions or molecules become essential to differentiating kidney epithelium, and many more probably await discovery. Some of these, like the original inductive signalling event described above, are almost certainly used to control cell behaviour in normal developing kidneys. Others, such as a sudden requirement for transferrin described below, are unlikely to regulate development in vivo, because the required molecule is always in plentiful supply. In most cases it is not yet possible to determine whether a potential checkpoint is used as such in vivo.

Differentiation of the stem cells results in the formation of two cell types; excretory epithelium and renal stroma. The intrinsic ability of the stem cell population to organise itself into these different developmental compartments is demonstrated most effectively by lithium-mediated induction, because the inducing stimulus is applied equally to all cells (in contrast to induction with heterologous tissues, which is selective for those cells at the interface). Nephrogenic mesenchymes treated with 15 m*M* Li^+ produce islands of cells undergoing epithelial differentiation amidst a sea of other cells expressing antigens characteristic of the stromogenic pathway (e.g. tenascin) [Aufderheide et al., 1987; Davies and Garrod, 1995; Davies and Garrod, in preparation]. Assignation to epitheliogenic or stromogenic pathways does not therefore require an external influence such as proximity to an inducing tissue. It might be determined by possible heterogeneity in the initial metanephrogenic mesenchyme (see above), or by a process such as lateral inhibition that creates two distinct phenotypes from an initially homogeneous population of induced stem cells.

The phenomenon of lateral inhibition has been studied most closely in *Drosophila* neurogenesis. Here a signalling system based on the transmembrane receptor *notch*, its membrane-bound ligand *delta*, and a feedback mechanism within each cell that regulates *delta* expression according to *notch* activity, results in the division of a population of cells into *delta*+ neurogenic cells and *notch*+ *delta*– epidermal cells [Simpson et al., 1992]. The vertebrate homologues of *notch*, mouse *notch 1* and *2*, are expressed in developing kidney, possibly in stem cells but certainly in condensed aggregates of cells that have just made the choice to follow the epithelial pathway [Weinmaster et al., 1991, 1992]. No information on the expression in kidneys of the vertebrate homologues of *delta* has been published, but it is tempting to speculate that the *notch/delta* mechanism might be involved in determining the fate of nephrogenic stem cells. There is therefore an urgent need to define the expression patterns of *notch* and *delta* in detail.

Recent experiments have suggested another candidate for determining fate – Wnt4. The exact time of its expression is not yet known, but Wnt4 appears to be activated after Pax2 and N-myc (i.e. after entry into the stem cell compartment) but before MET begins [Stark et al., 1994]. Transgenic animals carrying only non-functional mutant alleles of the gene fail to develop kidney tubules, showing that Wnt4 is required for nephrogenic MET. The stem cells also fail to express even their mutant Wnt4 gene, suggesting that wild-type Wnt4 is required for increased Wnt4 expression. This positive feedback, which might result in initially signal-dependent Wnt4 expression becoming self-sustaining, could be used to commit cells to a particular fate. It will be extremely interesting to see whether the kidneys of Wnt4 mutants contain only stem cells or whether instead all of the cells have become committed to the stromogenic pathway.

Many of the stem cell regulators discussed in the last section could also play a role in fate choice: interference with their function would be expected to result in disturbance of the balance between epitheliogenic and stromogenic cell types. Such a result has not yet been reported. Other molecules expressed by the stem cells and potentially involved in the processing of signals are 14-3-3-epsilon [McConnell et al., 1995] and SET [Kim et al., 1994], both of which are lost once the epithelial phenotype has become established. This temporal regulation suggests that they play an important developmental role.

The First Stage of Epithelial Differentiation – Aggregation

The first morphological sign that the MET will commence is the aggregation of still-mesenchymal cells into tight condensates that appear near the tips of the ureteric bud in vivo, and are located apparently randomly in lithium-mediated or transfilter spinal cord-mediated induction. The condensation process is preceded by an upregulation of the heparan sulphate proteoglycan, syndecan 1, and the cell adhesion molecule NCAM, which are, as far as we know, the first morphoregulatory molecules to be modulated in response to induction. Does their increased expression drive condensation?

NCAM is a calcium-independent homotypic cell-adhesion molecule expressed by a wide variety of neural and non-neural tissues [Rutishauser, 1992; Prieto and Crossin, 1995]. It could in principle increase the mutual adhesion of cells and thus force them to condense. However, function-blocking antibodies against NCAM fail to affect the formation of kidney tubules [Klein et al., 1988], and nephrogenesis proceeds normally in NCAM –/– mice [Cremer et al., 1994]. NCAM-mediated cell adhesion must either play no role in kidney tubulogenesis, or must be part of a highly redundant system. One possible molecule with which it could be redundant is a close relative of NCAM, DCC, which is expressed in early condensates [Poulsom and Hanby, pers. commun.].

Work on syndecan 1 has shown it to have a variety of functions, including cell-cell adhesion by homotypic binding [Fransson and Havmark, 1982], cell-matrix adhesion via interstitial collagens, fibronectin and tenascin [Koda et al., 1985; Saunders and Bernfield, 1988; Salmivirta et al., 1991] and presentation of growth factors to their high-affinity receptors [Rapraeger et al., 1991; Gallagher, 1994; Lyon et al., 1994]. The function of the complete syndecan 1 molecule has yet to be tested in this system, but the effects of removal of its sulphated glycosaminoglycan side chains, which are important to growth factor presentation and fibronectin binding, have been studied. In kidneys deprived of sulphated GAGs, collecting duct growth and branching is abolished but induction, condensation and epithelial differentiation of nephrogenic cells continues to take place [Davies et al., 1995]. Do NCAM and syndecan themselves form a redundant pair, or is condensation achieved by quite different molecules?

Acquisition of NCAM and syndecan marks the start of a series of changes in morphoregulatory molecule expression. The times of expression of a number of these molecules have now been determined; rather than appearing simultaneously, they are synthesised according to a highly reproducible sequence spread over some 18 h [Davies and Garrod, 1995] (fig. 3). At 24 h after induction, 6 h after the upregulation of NCAM and syndecan, and during the condensation phase, expression of extracellular matrix components changes in the condensates. Laminin B chains, which had been expressed at a low level in both the uninduced mesenchyme and the stem cells, are dramatically upregulated and are joined by collagen IV.

Around the time of condensation (the precise moment has yet to be determined), expression of the transcription factor WT1 increases dramatically. The genes coding for

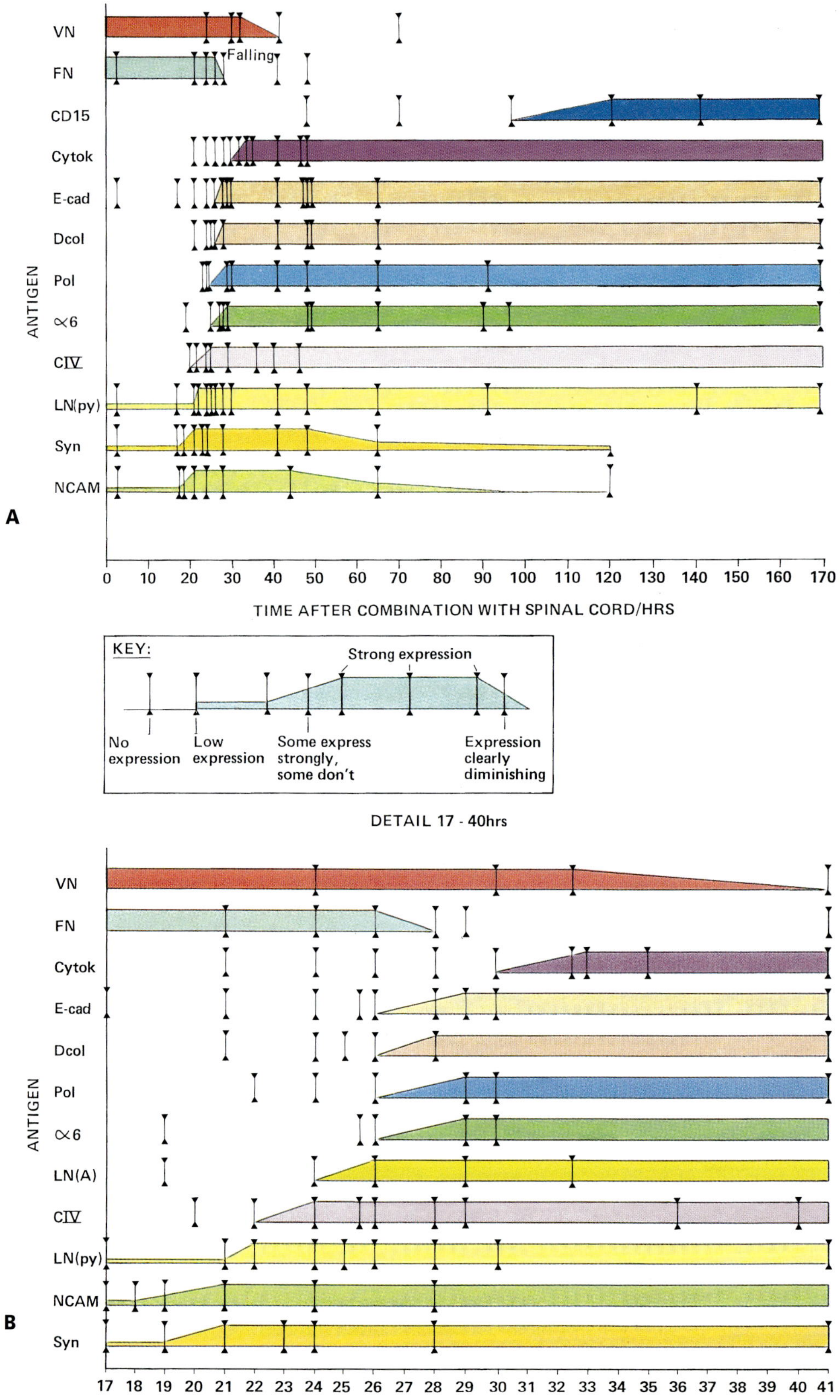
A
ANTIGEN
VN
FN
Falling
CD15
Cytok
E-cad
Dcol
Pol
∝6
CIV
LN(py)
Syn
NCAM
0 10 20 30 40 50 60 70 80 90 100 110 120 130 140 150 160 170
TIME AFTER COMBINATION WITH SPINAL CORD/HRS
KEY:
Strong expression
No expression
Low expression
Some express strongly, some don't
Expression clearly diminishing
DETAIL 17 - 40hrs
B
ANTIGEN
VN
FN
Cytok
E-cad
Dcol
Pol
∝6
LN(A)
CIV
LN(py)
NCAM
Syn
17 18 19 20 21 22 23 24 25 26 27 28 29 30 31 32 33 34 35 36 37 38 39 40 41

3

a number of growth factors present at the stem cell stage bear WT1-binding sequences in their promoter regions, arranged so that the binding of WT1 inhibits transcription; these factors include IGFII, the IGFI receptor and PDGF [Drummond et al., 1992; Gashler et al., 1992; Werner et al., 1994], WT1 may therefore be responsible for breaking the autocrine loops that drive stem cell proliferation.

A Checkpoint between Condensation and Epitheliogenesis Proper?

Two interesting observations made in culture suggest that there may be another checkpoint to development, located between the condensation stage and complete epithelial differentiation. Bard and Ross [1991] reported that addition of the multifunctional cytokine, LIF, to the medium of cultured developing kidneys resulted in a dramatic inhibition of nephrogenesis, although collecting duct morphogenesis proceeded normally. The early stages of nephron induction were unaffected by LIF, but development of nephrons stalled at the syndecan-expressing condensate stage [Bard, pers. commun.]. This stage of development also seemed to be significant in the experiments of Davies and Garrod [1995], who found that almost 50% of the condensates induced by Li^+-mediated induction differentiated no further. Taken together, these findings suggest that there may be another signal, inhibited by LIF and given inefficiently in the Li^+ induction system, that acts on the cells when condensation is complete. Its molecular basis is unknown (LIF is not naturally present in the kidney), but in other systems LIF is known to interact with IL6 signalling through gp130 [Boulton et al., 1994]. Perhaps exogenously added LIF is interfering with a similar signalling pathway in the condensates. (The receptor complex for cilliary neurotrophic factor (CNTF) also uses gp130 and the LIF receptor [Davis et al., 1993], but the normal development of kidneys in CNTF –/– transgenic animals [Masu et al., 1993] argues against a nephrogenic role for this growth factor.)

Fig. 3. Sequential expression of morphoregulatory molecules during nephrogenic MET. Coloured horizontal bars depict expression of each molecule during development, time elapsed since contact with an inducer being shown on the x-axis. Vertical black lines indicate times at which expression was assessed (immunocytochemically). **A** First week of development. **B** Period of MET to an expanded timescale. VN = Vimentin; FN = fibronectin; cytok = cytokeratin; E-cad = E-cadherin; Dcol = desmocollins; Pol = polarisation (of basement membrane components); α6 = α6-integrin; cIV = collagen IV; LN(A) = laminin A chain; LN(py) = laminin B chain; Syn = syndecan I. Data from Davies and Garrod [1995].

What might be the function of this putative checkpoint immediately before MET? One of the functions of condensation is the bringing together of a large enough number of epitheliogenic cells to be able to make the primitive nephric epithelium (see below and fig. 2). Possibly the LIF-inhibitable signal normally indicates to cells that their condensate has grown large enough to proceed; it has been known for some years that very small groups of metanephrogenic mesenchyme cannot be induced to form nephrons [Saxén, 1970]. The fact that condensates of Li^+-induced nephrogenic mesenchymes are rather smaller than those induced by living tissues [Davies and Garrod, 1995] fits well with this hypothesis, but there is no more evidence for it yet.

While they condense, cells acquire the CD71 transferrin receptor [Fleming and Jones, 1987] and gain a requirement for transferrin [Ekblom et al., 1981; Thesleff and Ekblom, 1984], or at least iron in a form deliverable to the cytoplasm [Landschulz et al., 1984]. This requirement is a potentially useful 'checkpoint' in vitro, though it is unlikely to be used in vivo due to the unrestricted availability of transferrin.

Acquisition of the Epithelial Phenotype

The stages of nephrogenesis described above, during which cells prepare for the mesenchyme-epithelium transition but remain completely mesenchymal, occupy the first 24 h after induction. Acquisition of an epithelial phenotype seems quite sudden in comparison, the process of MET being essentially complete 12 h later [all timings from Davies and Garrod, 1995]. The conversion of cells from condensed mesenchyme to simple epithelium is heralded by a number of simultaneous or nearly simultaneous changes in the expression of more morphoregulatory molecules.

From 26 h, cells in the differentiating condensates express laminin A chain in addition to the B chains they possessed already [Ekblom et al., 1990; Klein et al., 1990]. They also express α6β1 integrin hetrodimers, transmembrane receptors for laminin A that connect the cells to the basement membrane components they have begun to secrete [Sorokin et al., 1990]. As well as making new cell-substrate adhesion systems, the cells synthesise new cell-cell adhesion molecules characteristic of epithelia. From 28 h they express the calcium-dependent adhesion molecules E-cadherin and desmocollins, which are components of adhaerens junctions and desmosomes, respectively [Vestweber et al., 1985; Garrod and Fleming, 1990]; alpha-catenin, a cytoplasmic component of these junctions, is also acquired at about this time [Davies and Garrod, unpubl.]. All of the components mentioned in this paragraph are expressed around the membranes of apparently unpolarised cells, but once they are acquired the cells polarise rapidly; basement membrane components (e.g. laminin, collagen IV) become localised to the outside edge of the differentiating aggregate, and the cadherins are restricted to the lateral membranes.

A series of experiments by Ekblom and colleagues has demonstrated that the process of polarisation depends critically on cell-substrate interactions. Inhibition of the specific interaction between the laminin A chain and its transmembrane receptor $\alpha6\beta1$ integrin blocks the formation of polarised nephron epithelium in culture [Klein et al., 1988]. Similarly, blocking the interaction between laminin and nidogen, a protein that can link laminin and collagen IV molecules and which is present in the ECM surrounding differentiating nephrons, dramatically reduces the number of polarised epithelia that form [Ekblom et al., 1994].

Similar inhibition experiments on E-cadherin, the adhesive component of adhaerens junctions, have produced surprisingly little result [Vestweber et al., 1985]. Desmocollins are components of desmosomes, cell junctions that appear in at least a primitive state at this time [Garrod and Fleming, 1990]. Their presence at so plastic a stage of epithelial morphogenesis may be surprising in the light of models that consider them to be 'rivets' that lock cells together tightly [Staehelin, 1974], but lack of suitable reagents has left their developmental function so far untested.

The last stage in acquisition of a full epithelial phenotype is the exchange of a vimentin-based intermediate filament system for one composed of cytokeratins. This takes place surprisingly late: when they polarise and express desmosomal components, cells still contain vimentin and it is a full 6 h before cytokeratins appear and vimentin is lost.

Morphogenesis of the Newly Formed Epithelium

At this stage, the MET that is the subject of this review is essentially complete: cells that were once mesenchymal show all the characteristics of a normal simple epithelium. However, it is interesting to follow their development a little further, partly because of the complex morphogenetic movements involved in their maturation, and partly because a few of the cells quickly lose aspects of their hard-won epithelial phenotype.

Thin sections of the newly formed epithelium, and the diagrams that have been based on them, give the impression of an unbroken spherical epithelial sheet composed of polarised cells, with little or no lumen. The apparent symmetry of this structure is rapidly destroyed as cells rearrange themselves to create a deep curving invagination that characterises the 'comma-shape' stage of development, resulting in a curved tubule with a thin 'proximal' end and a thick 'distal' end (fig. 2). The stimulus for invagination and the mechanism that ensures that it takes place from only one point of the sphere remain utterly mysterious. The mechanical means by which the cell movements are generated are also unknown, although Saxén [1970] has proposed a mechanism of differential adhesion to explain both cell movement and proximo-distal polarity in the tubule. Further cell movements at the proximal pole create another cleft that will become the glomerular cavity. Together, the two clefts give the developing nephron the shape of a distorted letter 'S', hence the description 'S-shaped body' for this stage of development. In the S-shaped body, the proximo-distal polarity set up at or before the comma stage results in the differentiation of cell types specific to Bowman's capsule (part of the glomerulus) and the proximal and distal regions of the nephron itself.

Cells of the presumptive Bowman's capsule flatten considerably [Dorup and Maunsbach, 1982] and some differentiate into glomerular podocytes (fig. 4). Their differentiation is accompanied by changes in expression of transcription factors, including the loss of c-myc and the homeobox proteins Hox C9, LFB1 and LFB3 (all still expressed by the rest of the developing nephron) [Erselius et al., 1990; Lazzaro et al., 1990]. WT1 transcription continues [Armstrong et al., 1993]. As differentiation proceeds, the podocytes acquire a character that can be considered partially epithelial and partially mesenchymal. For example, like most epithelial cells, they express desmosomal components and $\alpha6$-integrin, but unusually they do not express cytokeratins or c-*met* [Garrod and Fleming, 1990; Holthofer et al., 1984; Sonnenburg et al., 1993; Woolf et al., 1994]. On the other hand, they do express $\alpha3$-integrin and vimentin, both more characteristic of mesenchyme [Korhonen et al., 1990, 1992; Holthofer et al., 1984]. Anatomically, the cell sheet formed by the podocytes is much more organised than a typical mesenchyme, but more leaky than a typical epithelium. The history of these cells therefore includes a MET followed almost at once by what may be regarded as a partial EMT!

The rest of the nephron remains typically epithelial but becomes regionally 'segmented'. Cell differentiation proceeds in a proximo-distal sequence, proximal markers (e.g. CD15) appearing 48 h before the distal ones (e.g. calbindin-D28k) [Davies, 1994]. The process is accompanied (timing is too uncertain to say more) by expression of a number of transcription factors (the homeobox proteins LFB1, Pax8) but loss of others, such as N-myc, MFH-1, WT1 and Pax2 [Lazzaro et al., 1990; Poleev et al., 1992; Mugrauer et al.,

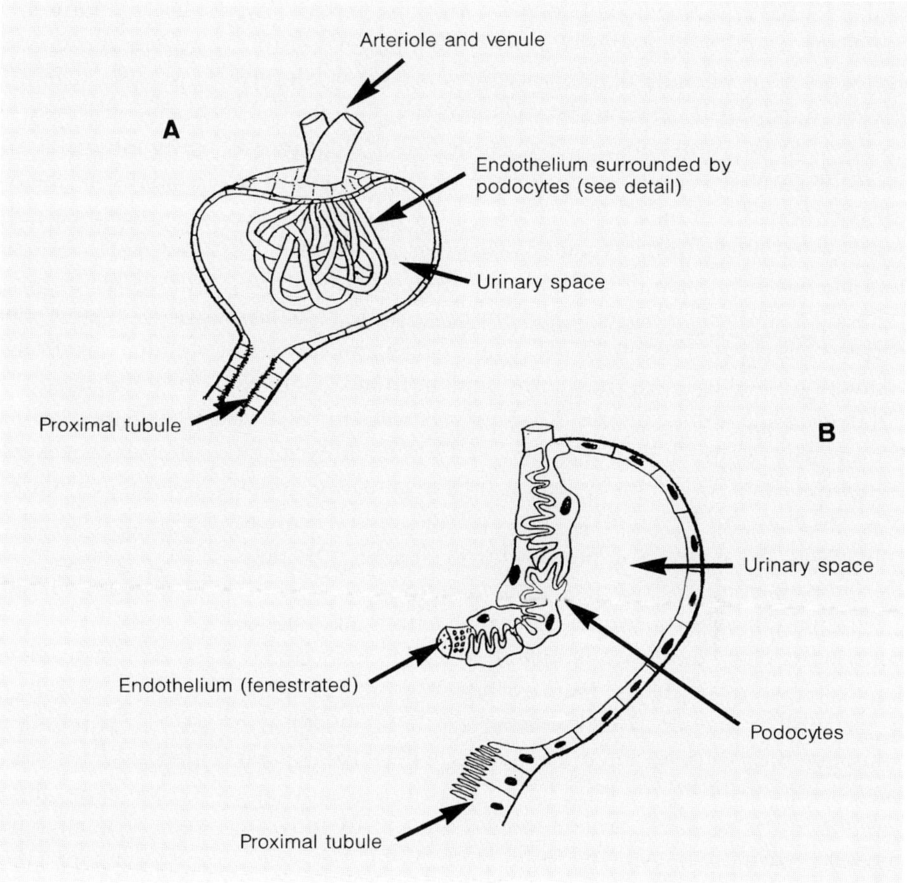

Fig. 4. Cell differentiation in the glomerulus. **A** The entire structure. **B** Detail to a larger scale, simplified by the inclusion of only one capillary. The podocytes surrounding the capillaries have characteristics of both epithelial and mesenchymal phenotypes.

1991; Miura et al., 1993; Armstrong et al., 1993; Dressler et al., 1992; Rothenpieler et al., 1993]. The functions of most of these proteins are unknown, but loss of Pax2 at this stage is important; while earlier expression of Pax2 is necessary for nephron formation to take place at all (see above), artificially prolonged Pax2 transcription in transgenic mice inhibits proper differentiation of the mature tubule and generates a pathological condition similar to nephrotic syndrome [Dressler et al., 1992].

The last major morphogenetic event is the joining of the newly formed excretory tubule to the collecting duct system, an event which requires the dissolution of two basement membranes and rearrangement of cell-cell contacts to create a continuous lumen.

A Pathological Failure of Nephrogenic MET: Wilms Tumour

One of the most common cancers of childhood, Wilms tumour, arises from a defect in the MET process described in this review. Tumours from different patients are not absolutely identical, but most have a characteristic triphasic histology consisting of uncontrolled growth of 'blastemal' cells surrounding small islands of epitheliogenic and stromogenic differentiation that parody normal kidney development.

The blastemal cells, which constitute the bulk of the tumour, do not express Pax2 [Dressler and Douglas, 1992]; they are therefore unlike normal nephrogenic stem cells. Most look like nephrogenic mesenchyme, though variants are more sarcomatoid (some Wilms tumours can be easily confused with rhabdomyosarcomas). The presence of epithelial differentiation (usually no more advanced than the comma/S stage) in typical tumours shows that the cells have not lost their capacity to undergo MET, though what induces them to do so in the depths of a tumour is a mystery.

About 60% of Wilms tumours are familial, the rest being sporadic. Study of the familial form has resulted in the identification of a tumour-suppressor gene that is mutated in approximately 15% of all Wilms tumours; the gene codes for the zinc finger protein WT-1 [for a recent review, see Coppes and Williams, 1994]. Another site, responsible for some (but not all) of the remaining cases, is closely linked to the IGFII gene on chromosome 11.

How do WT-1 mutations cause cancer? It is clear that a working copy of WT-1+ is necessary for kidneys to form at all (see above), so an affected foetus must inherit at least one normal allele from its parents; the loss of function of that allele takes place in affected cells after kidney development has begun [Haber and Housman, 1992]. As described above, expression of WT-1 normally increases dramatically as cells leave the stem population and begin to condense. The ability of WT-1 to suppress transcription of the growth factors IGFII, MK and PDGF encourages the hypothesis that WT-1 upregulation breaks autocrine loops that had been driving stem cell proliferation. A cell with no functional WT-1 genes could not break these loops; does this alone account for the neoplastic phenotype?

Wilms tumours do secrete abnormally high amounts of midkine [Uehara et al., 1992], PDGF [Gashler et al., 1992], nephroblast growth factor [Burrow and Wilson, 1993] and a gastrin-like peptide [Blackmore et al., 1994] which can stimulate mitosis in an autocrine manner. They also express high levels of IGFII message, but some of this is antisense and the levels of IGFII protein are not very different from those of normal kidneys [Baccarini et al., 1993; Yun et al., 1993]. The expression of other proteins, for example N-myc, is also abnormally strong in Wilms tumours [Nisen et al., 1986] but it is not clear whether this may be a cause or merely reflection of the large number of undifferentiated cells. If the entire explanation for the Wilms phenotype lies in the autocrine secretion of growth factors, it ought to be possible to produce a phenocopy of Wilms tumour by treating normal embryonic kidney stem cells with a cocktail of these growth factors in culture (Wilms tumour conditioned medium may be sufficient). As far as I know, this has not been achieved. In any case, it would be wrong to regard WT-1 simply as a repressor of growth. When the renal capacity of an adult animal is reduced, for example by removal of one kidney, there is a proliferative response in the contralateral kidney that, together with cell enlargement, increases the length (but not the number) of nephrons. One of the earliest events in this response is massive upregulation of WT-1 [Kushner et al., 1992].

The failure of most Wilms cells to differentiate is another aspect in need of explanation. Unless it arises from the autocrine loops referred to above (in which case experimental addition of pure growth factors should inhibit differentiation), it implies a direct role for WT-1 in the MET. Failure of maturation of the few tubules that do form in classical tumours is also unexplained – tissues distal to the glomerulus do not express WT-1 anyway, so why should its loss affect their development? Does their failure to mature imply that they require a normal glomerulus to set up the proximo-distal axis, or that the surrounding Wilms blastema cannot give them the stromal support they might require?

Questions proliferate. Answers await the results of numerous investigations currently being carried out. Among these are searches for other genes that the transcription factor regulates, which might when mutated cause the same cellular lesion as mutated WT1 itself and thereby account for the majority of Wilms tumours. Hopefully, the information gained will illuminate both normal development and tumorigenesis, and will suggest potential targets for therapeutic intervention.

Future Prospects for Study of Nephrogenic MET

Much of what has been learned about kidney development, especially at the molecular level, has been a fortuitous byproduct of experiments designed to establish the general embryonic expression pattern of a particular gene. Of the approximately 50 genes mentioned in this review, at least 30 fall into this category. Those of us interested in nephrogenic MET have therefore been very lucky to have benefited from such a windfall! However, obtaining a complete list of factors that can potentially regulate kidney development awaits the results of directed searches for genes modulated at precise stages of differentiation (defined by accurate timing or the use of checkpoints); at least two such searches are currently underway.

Enumeration of the players is only the first step in understanding their game. The function of each molecule involved can be determined only by careful inhibition/overexpression studies, but much circumstantial evidence will be obtained simply by establishing accurately the time of its expression. As well as ruling out some hypotheses (causes must precede effects), knowledge of relative timing of events can suggest a maximum number of successive gene activations in the causal chain between one event and another. Establishing which side of a checkpoint a gene is expressed will also give valuable clues about its developmental purpose.

Understanding the molecular basis for nephrogenic MET has implications not just for renal organogenesis, but for developmental biology as a whole. Examination of the step-by-step creation of an epithelium de novo from truly mesenchymal cells offers an opportunity to learn much about the epithelial state itself and the networks that main-

Acta Anatomica

Acta Anat 1996;156:202–216

L. Vićovac
J.D. Aplin

Department of Obstetrics and Gynaecology, and School of Biological Sciences, University of Manchester, UK

Epithelial-Mesenchymal Transition during Trophoblast Differentiation

Key Words
Trophoblast
Differentiation
Implantation
Placentation
Adhesion
Migration
Epithelial-mesenchymal transition

Abstract
Embryo implantation in higher primates is mediated by trophoblast: in the earliest phases by syncytiotrophoblast, then by both cytotrophoblast and syncytiotrophoblast. In the course of placentation three main trophoblast populations can be identified: cytotrophoblast stem cells and two differentiated derivative cell types: the syncytiotrophoblast and the extravillous cytotrophoblast. The syncytiotrophoblast remains mainly epithelial while the extravillous cytotrophoblast undergoes an epithelial-mesenchymal transition (EMT), initially forming multilayered cell columns and then, in human, infiltrating deeply the maternal decidual stroma and blood vessels. Finally, some infiltrating cells differentiate further to become giant cells of the placental bed and myometrium. During the course of these events the extravillous cytotrophoblast acquires a distinct phenotype, losing some typical epithelial components (e.g. E-cadherin, integrin $\alpha_6\beta_4$), but retaining others (e.g. cytokeratins). The signals that trigger this EMT are not well understood but its realisation is of critical importance for pregnancy success.

Introduction

The establishment and outcome of pregnancy depend critically on the formation of a zone of intimate physical contact and functional cooperation between the trophoblast and the uterus. Implantation in humans involves the embryo crossing the maternal uterine luminal epithelium to become embedded within the underlying endometrial stroma, eventually leading to the formation of a haemochorial placenta. In the course of placental development, an epithelial-mesenchymal transition (EMT) [Hay, 1995] takes place in which a large subpopulation of fetally-derived trophoblast cells escape from the conceptus to infiltrate the maternal interstitium and arteries (fig. 1a–d, 2). Other species including many primates and rodents also show interstitial implantation and a haemochorial placenta, but the human is unique in the extent to which mesenchymal transformation of trophoblast occurs [Pijnenborg et al., 1980; Aplin, 1991; Damsky et al., 1993; Aplin and Glasser, 1994; Cross et al., 1994; Pijnenborg, 1994]. Data presented in this paper will therefore concentrate on humans with only occasional reference to other species.

Trophoblast is extraembryonic fetal tissue derived from trophectoderm of the blastocyst, which in turn is the earliest

Abbreviations used in this paper:
CSF = Colony-stimulating factor; ECM = extracellular matrix; EGF = epidermal growth factor; EMT = epithelial-mesenchymal transition; GM-CSF = granulocyte-macrophage colony-stimulating factor; HB-EGF = heparin-binding epidermal growth factor; hCG = human chorionic gonadotrophin; IFN = interferon; IGF = insulin-like growth factor; IGF-BP1 = insulin-like growth factor binding protein 1; IL-1 = interleukin 1; LIF = leukaemia inhibitory factor; MHO = major histocompatibility complex; MMP = matrix metalloproteinase; NCAM = neural cell adhesion molecule; PA = plasminogen activator; PAI-1 = plasminogen activator inhibitor; PDGF = platelet-derived growth factor; PL = placental lactogen; PRL = prolactin; SF/HGF = scatter factor/hepatocyte growth factor; TGF = transforming growth factor; TIMP = tissue inhibitor of metalloproteinase; TNF = tumour necrosis factor; uPA = urokinase type plasminogen activator; VEGF = vascular endothelial growth factor.

KARGER
E-Mail karger@karger.ch
Fax + 41 61 306 12 34
http://www.karger.ch

0001–5180/96/1563–0202$12.00/0

John D. Aplin, PhD
Research Floor
St. Mary's Hospital
Manchester M13 0JH (UK)

Nisen, P.D., K.A. Zimmerman, S.V. Cotter, F. Gilbert, F.W. Alt (1986) Enhanced expression of the N-myc gene in Wilms' tumors. Cancer Res *46:* 6217–6222.
Orr-Urtreger, A., D. Givol, A. Yayon, P. Lonai (1991) Developmental expression of two murine fibroblast growth factor receptors, flg and bek. Development *113:* 1419–1434.
Phelps, D.E., G.R. Dressler (1993) Aberrant expression of Pax-2 in Danforth's short tail (Sd) mice. Dev Biol *157:* 251–258.
Poleev, A., H. Fickenscher, S. Mundlos, A. Winterpacht, B. Zabel, A. Fidler, P. Gruss, D. Plachov (1992) Pax 8, a human paired box gene: Isolation and expression in developing thyroid, kidney and Wilm's tumours. Development *116:* 611–623.
Prieto, A.L., K.L. Crossin (1995) Cell-cell adhesion molecule in epithelial-mesenchymal transformations. Acta Anat *154:* 21–33.
Rapraeger, A.C., A. Krufka, B.B. Olwin (1991) Requirement for heparan sulphate for bFGF mediated fibroblast growth and myoblast differentiation. Science *252:* 1705–1708.
Rivera-Madrid, R., P. Marinko, Y. Chartier, Y. Meyer (1993) Nucleotide sequence of an *Arabidopsis thaliana* cDNA clone encoding a homolog to a suppressor of Wilms' tumour. Plant Physiol *102:* 329–330.
Rothenpieler, U.W., G.R. Dressler (1993) Pax-2 is required for mesenchyme-to-epithelium conversion during kidney development. Development *119:* 711–720.
Rutishauser, U. (1992) NCAM and its polysialic acid moiety: a mechanism for pull/push regulation of cell interactions during development. Development *suppl (1992):* 99–104.
Salmivirta, M., K. Elenius, S. Vainio, U. Hofer, R. Chiquet-Ehrissman, I. Thesleff, M. Jalkanen (1991) Syndecan from tooth mesenchyme binds tenascin. J Biol Chem *266:* 7733–7739.
Santos, O.F.P., E.J.G. Baros, X.M. Yang, K. Malsumoto, T. Nakamura, M. Park, S.K. Nigam (1994) Involvement of hepatocyte growth factor in kidney development. Dev Biol *163:* 525–529.
Sariola, H., K. Holm, S. Henke-Fahle (1988) Early innervation of the metanephric kidney. Development *104:* 589–599.
Sariola, H., M. Saarma, K. Sainio, U. Arumäe, J. Palgi, A. Vaahtokari, I. Thesleff, A. Karavanov (1991) Dependence of kidney morphogenesis on the expression of nerve growth factor receptor. Science *254:* 571–573.
Saunders, S., M. Bernfield (1988) Cell surface proteoglycan binds mouse mammary epithelial cells to fibronectin and behaves as a receptor for interstitial matrix. J Cell Biol *106:* 423–430.
Saxén, L. (1970) Failure to demonstrate tubule induction in a heterologous mesenchyme. Dev Biol *23:* 511–523.
Saxén, L. (1980) Mechanism of morphogenetic tissue interactions: The message of transfilter experiments; in McKinnell RG, MA DiBerardino, M Blumenfield, RD Bergad (eds): Results and Problems in Cell Differentiation, vol 11: Differentiation and Neoplasia. Berlin, Springer, pp 147–154.
Saxén, L. (1987) Organogenesis of the Kidney. Cambridge, Cambridge University Press.
Saxén, L., E. Lehtonen (1978) Transfilter induction of kidney tubules as a function of the extent and duration of intercellular contacts. J Embryol Exp Morphol *47:* 97–109.
Saxén, L., E. Lehtonen, M. Karkinen-Jääskelainen, A. Nordling, J. Wartiovaara (1976) Are morphogenetic tissue interactions mediated by transmissible signal substances or through cell contacts? Nature *259:* 662–663.
Saxén, L., J. Salonen, P. Ekblom, S. Norling (1983) DNA synthesis and cell generation cycle during determination and differentiation of the metanephric mesenchyme. Dev Biol *98:* 130–138.
Schmid, P., A. Lorenz, H. Hameister, M. Montenarh (1991) Expression of p53 during mouse embryogenesis. Development *113:* 857–865.
Schuchardt, A., V. D'Agati, L. Larsson-Blomberg, F. Constantini, V. Pachnis (1994) Defects in the kidney and enteric nervous system of mice lacking the tyrosine kinase receptor Ret. Nature *367:* 380–383.
Sharpe, P.T., J.R. Miller, E.P. Evans, M.D. Burtenshaw, S.J. Gaunt (1988) Isolation and expression of a new mouse homoeobox gene. Development *102:* 397–407.
Simpson, P., M. Bourouis, P. Heitcler, L. Rule, M. Haenlin, P. Ramain (1992) Delta, notch and shaggy: Elements of a lateral signalling pathway in *Drosophila.* Cold Spring Harbor Symp Quant Biol *57:* 391–400.
Slack, J.M. (1993) Embryonic induction. Mech Dev *41:* 91–107.
Sonnenberg, E., D. Meyer, K.M. Weidner, C. Birchmeier (1993) Scatter factor/hepatocyte growth factor, the c-*met* tyrosine kinase, can mediate a signal exchange between mesenchyme and epithelium during mouse development. J Cell Biol *123:* 223–225.
Sorokin, L., A. Sonnenberg, M. Aumailley, R. Timpl, P. Ekblom (1990) Recognition of the laminin E8 cell-binding site by an integrin possessing the α6 subunit is essential for epithelial polarisation in developing kidney tubules. J Cell Biol *111:* 1265–1273.
Staehlin, A. (1974) Structure and function of intercellular junctions. Int Rev Cytol *39:* 191–283.
Stanton, B.R., A.S. Perkins, L. Tessarollo, D.A. Sassoon, L.F. Parada (1992) Loss of N-*myc* function results in embryonic lethality and failure of the epithelial compartment of the embryo to develop. Genes Dev *6:* 2235–2247.
Stark, K., S. Vainio, G. Vassileva, A.P. McMahon (1994) Epithelial transformation of metanephric mesenchyme in the developing kidney regulated by Wnt 4. Nature *372:* 679–683.
Thesleff, I., P. Ekblom (1984) Role of transferrin in branching morphogenesis, growth and differentiation of the embryonic kidney. J Embryol Exp Morphol *82:* 147–161.
Tsarfaty, I., S. Rong, J.H. Resau, S. Rulong, P.P. da-Silva, G.F. Vande Woude (1994) The met proto-oncogene mesenchymal to epithelial cell conversion. Science *263:* 98–101.
Tsutsui, J., K. Kadomatsu, S. Matsubara, A. Nakagawara, M. Hamonoue, S. Takao, H. Shimazu, Y. Ohi, T. Muramatsu (1994) A new family of heparin-binding growth/differentiation factors: Increased midkine expression in Wilms tumor and other human carcinomas. Cancer Res *53:* 1281–1285.
Uehara, K., S. Matsubara, K. Kadomatsu, J. Tsutsui, T. Muramatsu (1992) Genomic structure of human midkine (MK), a retinoic acid-response growth/differentiation factor. J Biochem *111:* 563–567.
Unsworth, B., C. Grobstein (1970) Induction of kidney tubules in mouse metanephrogenic mesenchyme by various embryonic mesodermal tissues. Dev Biol *21:* 547–556.
Vestweber, D., R. Kemler, P. Ekblom (1985) Cell adhesion molecule uvomorulin during kidney development. Dev Biol *112:* 213–221.
Wall, N.A., C.M. Jones, B.L. Hogan, C.V. Wright (1992) Expression and modification of Hox 2.1 protein in mouse embryos. Mech Dev *92:* 111–120.
Wartiovaara, J., S. Nordling, E. Lehtonen, L. Saxén (1974) Transfilter induction of kidney tubules: Correlation with cytoplasmic penetration into Nucleopore filters. J Embryol Exp Morphol *31:* 667–682.
Weinmaster, G., V.J. Roberts, G. Lemke (1991) A homolog of *Drosophila* Notch expressed during mammalian development. Development *113:* 199–205.
Weinmaster, G., V.J. Roberts, G. Lemke (1992) Notch2: A second mammalian Notch gene. Development *116:* 931–941.
Weller, A., L. Sorokin, E.M. Illgen, P. Ekblom (1991) Development and growth of mouse embryonic kidney in organ culture and modulation of development by soluble growth factor. Dev Biol *144:* 248–261.
Werner, H., F.J. Rauscher, 3rd, V.P. Sukhatme, I.A. Drummond, C.T. Roberts, Jr., D. LeRoith (1994) Transcriptional repression of the insulin-like growth factor I receptor (IGFI-R) gene by the tumor suppressor WT1 involves binding to sequences both upstream and downstream of the IGFI-R gene transcription start site. J Biol Chem *269:* 12577–12582.
Weston, J.A. (1963) A radioautographic analysis of the migration and localisation of trunk neural crest cells in the chick. Dev Biol *6:* 279–310.
Wilkinson, D.G., J.A. Bailes, A.P. McMahon (1987) Expression of the protooncogene int-1 is restricted to specific neuronal cells in the developing mouse embryo. Cell *50:* 79–88.
Woolf, A.S., M. Kolatsi, P. Hardman, E. Andermacher, C. Moorby, G. Fine, P.S. Jat, M.D. Noble, E. Gherardi (1994) Roles of hepatocyte growth factor/scatter factor (HGF/SF) and the met receptor in the early development of the metanephros. J Cell Biol *128:* 171–184.
Yun, K., A.T. Molenaar, A.M. Fiedler, A.J. Mark, M.R. Eccles, D.M. Becroft, A.E. Reeve (1993) Insulin-like growth factor II messenger RNA expression in Wilms tumour, nephrogenic rest and kidney. Lab Invest *69:* 603–615.

Ekblom, P., I. Thesleff, A. Miettinen, L. Saxén (1981) Organogenesis in a defined medium supplemented with transferrin. Cell Different *10:* 281–288.
Ekblom, P., I. Thesleff, L. Saxén, A. Miettinen, R. Timpl (1983) Transferrin as a foetal growth factor: Acquisition of responsiveness related to embryonic induction. Proc Natl Acad Sci USA *80:* 2651–2655.
Erselius, J.R., M.D. Goulding, P. Gruss (1990) Structure and expression pattern of the murine Hox 3.2 gene. Development *110:* 629–642.
Fleming, S., D.B. Jones (1987) Immunocytochemical evidence for transferrin-dependent proliferation during renal tubulogenesis. J Anat *153:* 191–201.
Fransson, L.A., B. Havmark (1982) Structural requirement for heparan sulphate self association. Carbohyd Res *11:* 135–144.
Gallagher, J.T. (1994) Heparan sulphates as membrane receptors for the fibroblast growth factors. Eur J Clin Chem Clin Biochem *1:* 1201–1208.
Garrod, D.R., S. Fleming (1990) Early expression of desmosomal components during kidney tubule morphogenesis in human and murine embryos. Development *108:* 313–321.
Garvin, A.J., G.G. Re, B.I. Tarnowski, D.J. Hazen Martin, D.A. Sens (1993) The G401 cell line, utilized for studies of chromosomal changes in Wilms tumor, is derived from rhabdoid tumour of the kidney. Am J Pathol *142:* 375–380.
Gashler, A.L., D.T. Bonthron, S.L. Madden, F.J. Rauscher, 3rd, T. Collins, V.P. Sukhatme (1992) Human platelet-derived growth factor A chain is transcriptionally repressed by the Wilms tumor suppressor WT1. Proc Natl Acad Sci USA *89:* 10941–10988.
Gaunt, S.J. (1987) Homoeobox gene Hox 1.5 expression in mouse embryos: Earliest detection by in situ hybridisation is during gastrulation. Development *101:* 51–60.
Gaunt, S.J. (1988) Mouse homoeobox gene transcripts occupy different but overlapping domains in embryonic germ layers and organs: A comparison of Hox 3.1 and Hox 1.5. Development *103:* 135–144.
Gessler, M., A. Poustka, W. Cavanee, R.L. Neve, S.H. Orkin, G.A. Bruns (1990) Homozygous deletion in Wilms tumours of a zinc finger gene identified by chromosome jumping. Nature *343:* 774–778.
Gherardi, E., M. Sharpe, K. Lane, A. Sirulnik, M. Stoker (1993) Hepatocyte growth factor/scatter factor (HGF/SF), the c-met receptor and the behaviour of epithelial cells. Symp Soc Exp Biol *47:* 163–181.
Grobstein, C. (1953) Inductive epithelio-mesenchymal interactions in cultured organ rudiments of the mouse. Science *118:* 52–55.
Grobstein, C. (1955) Inductive interactions in the development of the mouse metanephros. J Exp Zool *130:* 319–340.
Haber, D.A., D.E. Housman (1992) The genetics of Wilms tumor. Adv Cancer Res *59:* 41–68.
Herzlinger, D., R. Abramson, D. Cohen (1993) Phenotypic conversions in renal development. J Cell Sci *(suppl) 17:* 61–64.
Holland, P.N., B.L. Hogan (1988) Spatially restricted patterns of expression of the homeobox-containing gene Hox 2.1 during mouse embryogenesis. Development *102:* 159–174.
Holthofer, H., A. Miettinen, V.P. Lehto, E. Lethonen, I. Virtanen (1984) Expression of vimentin and cytokeratin types of intermediate filament proteins in developing and adult human kidneys. Lab Invest *50:* 552–559.
Kanwar, Y.S., J. Wada (1994) Insulin-like growth factor I receptor in metanephric development. Contrib Nephrol. Basel, Karger, vol 107, pp 168–173.
Karp, S.L., A. Oritz-Arduan, S. Li, E.G. Neilson (1994) Epithelial differentiation of metanephric mesenchymal cells after stimulation with hepatocyte growth factor or embryonic spinal cord. Proc Natl Acad Sci USA *91:* 5286–5290.
Kim, E.G., M.E. Choi, B.J. Ballermann (1994) Spatially restricted expression of SET mRMA in developing rat kidney. Am J Physiol *260:* F135–161.
Kissane, J.M. (1983) Congenital malformations; in Heptinstall E (ed): Pathology of the Kidney, ed 3. London, Churchill Livingstone, pp 83–140.
Klein, G., M. Ekblom, L. Fecher, R. Timpl, P. Ekblom (1990) Differential expression of laminin A and B chains during development of embryonic mouse organs. Development *110:* 823–837.
Klein, G., M. Langegger, R. Timpl, P. Ekblom (1988) Role of laminin A chain in the development of epithelial cell polarity. Cell *55:* 331–341.
Klein, G., M. Langegger, C. Goridis, P. Ekblom (1988) Neural cell adhesion molecule during induction and development of the kidney. Development *102.* 749–761.
Koch, W.G., C. Grobstein (1963) Transmission of radioisotopically labelled materials during embryonic induction in vitro. Dev Biol *7:* 303–323.
Koda, J.E., A. Rapraeger, M. Bernfield (1985) Heparan sulphate proteoglucans from mouse mammary epithelial cells: cell surface proteoglycan as a receptor for interstitial collagens. J Biol Chem *260:* 8157–8162.
Korhonen, M., L. Laitenen, J. Ylanne, V.S. Gould, I. Virtanen (1992) Integrins in developing, normal and pathological human kidney. Kidney Int *41:* 641–644.
Korhonen, M., J. Ylanne, L. Laitinen, I. Virtanen (1990) The $\alpha1-\alpha6$ subunits of integrins are characteristically expressed in distinct segments of developing and adult human nephron. J Cell Biol *111:* 1245–1254.
Koseki, C., D. Herzlinger, Q. al-Awqati (1992) Apoptosis in metanephric development. J Cell Biol *119:* 1327–1333.
Kushner, L., A. Cozzolino, J. Sherman, M.A. Rich (1992) Expression of a Wilms tumour gene in porcine kidney during compensatory renal growth. J Urol *148:* 555–558.
Landshulz, W., I. Thesleff, P. Ekblom (1984) A lipophilic iron chelator can replace transferrin as a stimulator of cell proliferation and differentiation. J Cell Biol *98:* 596–601.
Lazzaro, D., V. DeSimone, L. DeMagistris, E. Lehtonen, R. Cortese (1992) LFB1 and LFB3 homeoproteins are sequentially expressed during kidney development. Development *114:* 469–479.
LeDouarin, N.M., M.-A.M. Teillet (1974) Experimental analysis of the migration and differentiation of neuroblasts of the autonomic nervous system and of neurectodermal mesenchymal derivatives using a biological cell marking technique. Dev Biol *41:* 162–184.
Lee, K.F., E. Li, L.J. Huber, S.C. Landis, A.H. Sharpe, M.V. Chao, R. Jaenisch (1992) Targeted disruption of the gene encoding the low affinity NGF receptor p75 leads to defects in the peripheral sensory nervous system. Cell *69:* 737–749.
Lehtonen, E. (1976) Transmission of signals in embryonic kidney. Med Biol *54:* 108–128.
Lemoine, N.R., C.M. Hughes, J.K. Cowell (1992) Aberrant expression of the tumour suppressor gene p53 is very frequent in Wilms tumor. J Pathol *168:* 237–242.
Liu, Z.Z., J. Wada, K. Alvares, A. Kumar, E.I. Wallner, Y.S. Kanwar (1993) Distribution and relevance of insulin-like growth factor I receptor in metanephric development. Kidney Int *44:* 1242–1250.
Lombard, M.-N., C. Grobstein (1969) Activity in various embryonic and postembryonic sources for induction of kidney tubules. Dev Biol *19:* 41–51.
Lyon, M., J.A. Deakin, K. Mizumo, T. Nakamura, J.L. Gallagher (1994) Interaction of hepatocyte growth factor with heparan sulphate. J Biol Chem *269:* 11216–11223.
Masu, Y., E. Wolf, B. Holtmann, M. Sendtner, G. Brem, H. Thoenen (1993) Disruption of the CNTF gene results in motor neuron degeneration. Nature *165:* 27–32.
Matsell, J. (1993) Abstract. J Am Soc Nephrol *4:* 471.
McConnell, J.E., J.F. Armstrong, P.E. Hodges, J.B.L. Bard (1995) Expression of 14-3-3 epsilon during metanephric development. Dev Biol *169:* 218–228.
McMahon, A.P., A. Bradley (1990) The Wnt-1 (int1) proto-oncogene is required for the development of a large region of the mouse brain. Cell *62:* 1073–1085.
Miura, M., A. Wanaka, M. Tokyama, K. Tanaka (1993) MFH-1, a new member of the fork head domain family, is expressed in developing mesenchyme. FEBS Lett *326:* 171–176.
Mugrauer, G., P. Ekblom (1991) Contrasting expression patterns of 3 members of the myc family of protooncogenes in the developing and adult mouse kidney. J Cell Biol *112:* 13–25.
Muramatsu, H., H. Shirakama, S. Yonezawa, H. Murata, T. Muramatsu (1993) Midkine, a retinoic acid-inducible growth factor/differentiation factor: Immunochemical evidence for the function and distribution. Dev Biol *159:* 392–402.

tain it. Indeed, greater knowledge of this MET and the EMTs described elsewhere in this volume will help to define just how much, or how little, separates the mesenchymal and epithelial categories into which we have traditionally divided all somatic cells.

The Kidney Development Database

The expression during kidney development of a large number of genes is now summarised, together with other relevant information, in the Kidney Development Database. A hypertext version of the database can be accessed by a suitable web browser at <*http://mbisg2.sbc.man.ac.uk/kidbase/kidhome.html*> and at <*http://www.ana.ed.ac.uk./anatomy/kidbase/kidhome.html*>.

Acknowledgements

I would like to thank Prof. David Garrod for his comments on the manuscript, and the Cancer Research Campaign, Royal Society and Medical Research Council for their support of my work.

References

Armstrong, J.F., K. Pritchard-Jones, W. Bickmore, N.D. Hastie, J.B. Bard (1993) The expression of the Wilms tumour gene, *WT*1, in the developing mammalian embryo. Mech Dev *40:* 85–97.

Auerbach, R., C. Grobstein (1958) Inductive interaction of embryonic tissues after dissociation and reaggregation. Exp Cell Res *15:* 384–392.

Aufderheide, F., R. Chiquet-Ehrissman, P. Ekblom (1987) Epithelial-mesenchymal interactions in the developing kidney lead to expression of tenascin in the mesenchyme. J Cell Biol *105:* 599–608.

Baccarini, P., M. Fiorentino, A. D'Errico, A.M. Mancini, W.F. Grigioni (1993) Detection of antisense transcripts of the insulin-like growth factor 2 gene in Wilms tumor. Am J Pathol *143:* 1535–1542.

Bard, J.B.L. (1992) The development of the mouse kidney: Embryogenesis writ small. Curr Opin Genet Dev *2:* 589–595.

Bard, J.B.L., J.E. McConnell, J.A. Davies (1994) Towards a genetic basis for kidney development. Mech Dev *48:* 3–11.

Bard, J.B.L., A. Ross (1991) LIF, the ES-cell inhibition factor, reversibly blocks nephrogenesis in cultures mouse kidney rudiments. Development *113:* 193–198.

Birch, N. (1991) Lithium and the Cell. New York, Academic Press.

Blackmore, M., E. Doherty, J.E. Manning, B.H. Hirst (1994) Autocrine stimulation of human renal Wilms tumor G401 cells by a gastrin-like peptide. Int J Cancer *57:* 385–391.

Boulton, T.G., N. Stahl, G.D. Yancopoulos (1994) Ciliary neurotrophic factor/leukaemia inhibitory factor/interleukin 6/oncostatin M family of cytokines induce tyrosine phosphorylation of a common set of proteins overlapping those induced by other cytokines and growth factors. J Biol Chem *269:* 11648–11655.

Burrow, C.R., P.D. Wilson (1993) A putative Wilms Tumour-secreted growth factor activity required for primary culture of human nephroblasts. Proc Natl Acad Sci USA *90:* 6066–6090.

Cardia, A.F., J. Hu, J. Crosby, P.A. Lalley, D. Noden, J.H. Nadeau, C.V. Wright (1992) Mox-1 and Mox-2 define a novel homeobox gene subfamily and are differentially expressed during early mesodermal patterning in mouse embryos. Development *116:* 1123–1136.

Coppes, M.J., B.R. Williams (1994) The molecular genetics of Wilms' tumor. Cancer Invest *12:* 57–65.

Cremer, H., R. Lange, A. Christoph, M. Plomann, G. Vopper, J. Roes, R. Brown, S. Baldwin, P. Kearmer, S. Scheff, D. Barthels, K. Rajewsky, W. Wille (1994) Inactivation of the N-CAM gene in mice results in size reduction of the olfactory bulb and defects bulb and defects in spatial learning. Nature *367:* 455–458.

Davies, J.A. (1993) How to build a kidney. Semin Cell Biol *4:* 213–219.

Davies, J.A. (1994) Control of Calbindin D-28K expression in developing mouse kidney. Dev Dynam *199:* 45–51.

Davies, J.A., A.W. Brandli (1996) The Kidney Development Database. World Wide Web - address *http://mbisg2.sbc.man.ac.uk/kidbase/kidhome.html.*

Davies, J.A., D.R. Garrod (1995) Induction of early stages of kidney tubule differentiation by lithium ions. Dev Biol *167:* 50–60.

Davies, J.A., M. Lyon, J. Gallagher, D.R.G. Garrod (1995) Sulphated proteoglycan is required for collecting duct growth and branching but not nephron formation during kidney development. Development *121:* 1507–1517.

Davis, S., T.H. Aldrich, N. Stahl, L. Pan, T. Taga, T. Kishimoto, N.Y. Ip, G.D. Yancopoulos (1993) LIFR beat and gp 130 as heterodimerizing signal transducers of the tripartite CNTF receptor. Science *360:* 1805–1808.

Dorup, J., A.B. Maunsbach (1982) The ultrastructural development of distal nephron segments in the human fetal kidney. Anat Embryol *164:* 19–41.

Dressler, G.R. (1994) Developmental regulation and oncogenesis in the kidney; in Liotta LA (chairman): Molecular Genetics of Tumour Progression and Metastasis: A Special Conference of the American Association for Cancer Research.

Dressler, G.R., E.C. Douglas (1992) Pax2 is a DNA-binding protein expressed in embryonic kidney and Wilms tumor. Proc Natl Acad Sci USA *89:* 1179–1183.

Dressler, G.R., U.W. Rothenpieler, L.T. Patterson, L. Williams-Simons, H. Westphal (1992) Deregulation of Pax-2 expression in transgenic mice generates severe kidney abnormalities. Nature *362:* 65–67.

Drummond, I.A., S.L. Madden, P. Rohuer-Nutter, G.I. Bell, V.P. Sukhatme, F.J. Rauscher, 3rd (1992) Repression of the insulin-like growth factor II gene by the Wilms tumor suppressor WT1. Science *257:* 674–678.

Duband, J.L., S. Dufour, K. Hatta, M. Takeichi, G.M. Edelman, J.P. Thiery (1987) Adhesion molecules during somitogenesis in the avian embryo. J Cell Biol *104:* 1361–1374.

Dunhill, M.S., W. Haley (1972) Some observations on the quantitative anatomy of the kidney. J Pathol *110:* 113–121.

Durbeej, M., S. Söderström, T. Ebendal, C. Birchmeier, P. Ekblom (1993) Differential expression of neurotrophin receptors during renal development. Development *119:* 977–989.

Ekblom, P., M. Ekblom, L. Fecher, G. Kelin, H.-Y. Zhang, Y. Kaebuya, M.-L. Chu, V. Mayer, R. Timpl (1994) Role of mesenchymal nidogen for epithelial morphogenesis in vitro. Development *120:* 1997–2002.

Ekblom, P., G. Klein, M. Ekblom, L. Sorokin (1990) Laminin isoforms and their receptors in the developing kidney. Am J Kid Dis *17:* 603–605.

Ekblom, P., G. Klein, M. Ekblom, L. Sorokin (1991) Laminin isoforms and their receptors in the developing kidney. Am J Kid Dis *17:* 603–605.

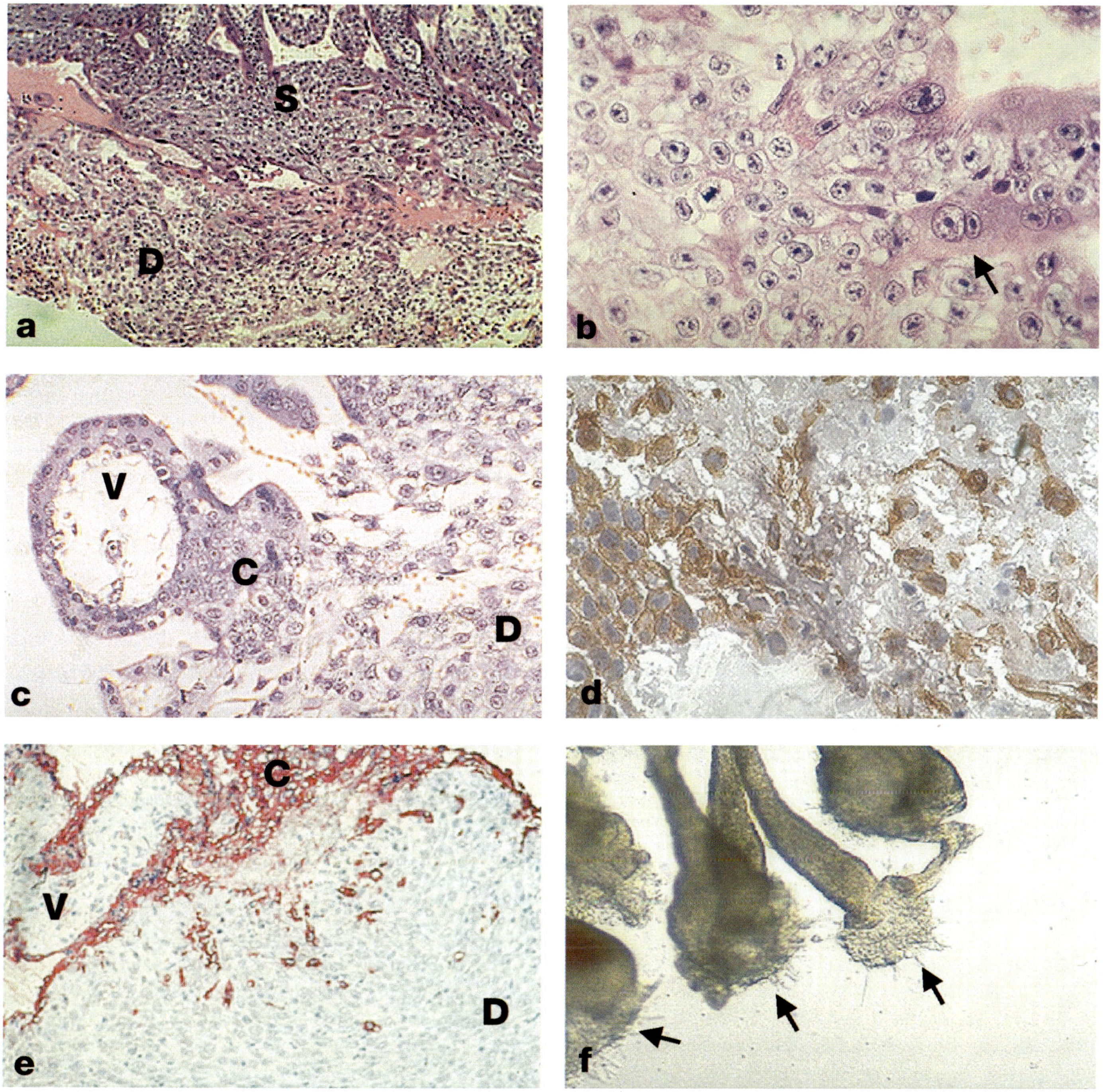

Fig. 1. **a** Human implantation site at day 13 post-ovulation. At the top of the field are spaces that have developed from lacunae within the early syncytium. At their borders are villi invested with early mesenchyme. S marks the trophoblastic shell, which contains both cyto- and syncytiotrophoblast. D indicates the maternal decidua which also contains a significant population of infiltrating trophoblast. **b** Higher power view of the same specimen. The blood spaces are lined by a mixture of syncytio- (arrow) and cytotrophoblast (visible at the top). The shell is seen to contain both types of trophoblast. **c** A small anchoring villus of late first trimester. V = Villous stroma; C = cytotrophoblast column; D = decidua. **d** First-trimester placental bed specimen stained with anti-cytokeratin antibody to visualise the infiltrating cytotrophoblast population. **e** Coculture experiment in which floating villus tissue (V) was maintained in contact with parietal (non-placental bed) decidua (D). Staining is for cytokeratin. The 48-hour time point is shown; de novo formation of columns (C) has occurred and there is significant infiltration of the decidua by cytotrophoblast. **f** First-trimester villous tissue explanted into a collagen gel. Bulbous cytotrophoblast columns are seen developing at the tips of the villi and elongated cells extend into the gel (arrows). **a**–**c** HE; **d**, **e** immunoperoxidase; **f** no staining.

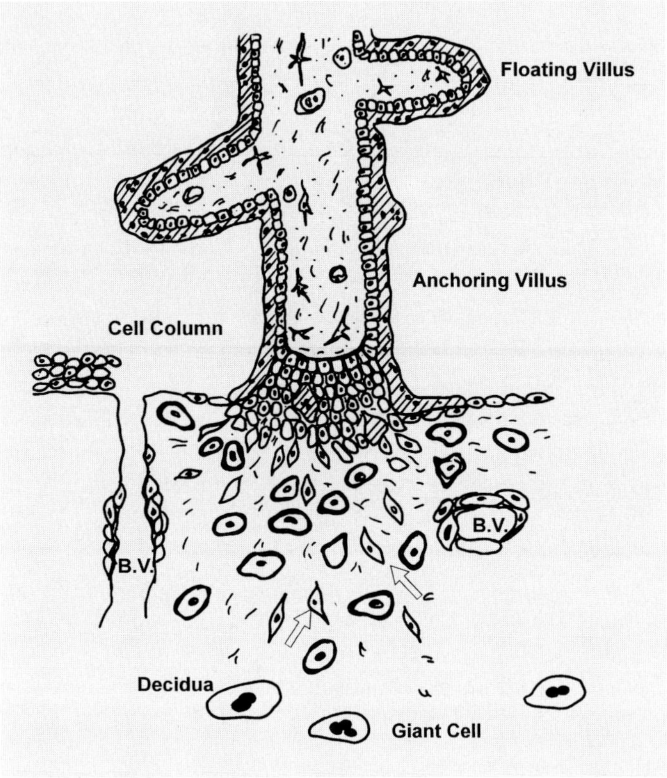

Fig. 2. Floating and anchoring villi in first trimester. A mesenchymal villus is shown with two floating branches into the intervillous space. These are covered with an epithelial-type stem cytotrophoblast monolayer outside which is a continuous epithelial syncytium. Cytotrophoblast proliferation occurring at the site of anchorage to decidua leads to the generation of a column from which trophoblast undergo an EMT, single cells elongating and detaching (arrows) to infiltrate the maternal decidual interstitium and arteries (B.V.). The infiltrating cells eventually differentiate to become placental bed giant cells, situated in the decidua or inner myometrium.

epithelium. Apical membrane interaction between trophectoderm and uterine luminal epithelium leads to penetration of the maternal basement membrane [Schlafke and Enders, 1975; Denker, 1990, 1993] and is believed to be possible only during a receptive phase a few days in duration [Psychoyos, 1988; Bergh and Navot, 1992; Aplin, 1996]. There follows a short period of intense invasivity during which the conceptus becomes established within the maternal stroma. Implantation is followed by a period of rapid trophoblast proliferation and placental morphogenesis [Enders et al., 1996], initially greatly exceeding the proliferative activity in the developing germ disc. Normal development of the embryo presumably depends on this early phase of placentation. Early syncytiotrophoblast and cytotrophoblast can be recognised during the intrusion phase of implantation [Hertig et al., 1956; Enders, 1991]. These early phases of implantation have been reviewed in detail elsewhere [Enders, 1991; Denker, 1993] and are not further described here. Subsequently, with progression of placentation, trophoblast can be found forming a variety of different structures, e.g. placental villi, fetal membranes, trophoblast shell (in first trimester), cell islands and anchoring villi [Benirschke and Kaufmann, 1990]. They all arise from villous cytotrophoblast stem cells which either fuse to produce the (epithelial) syncytiotrophoblast that covers placental villi [Hoshina et al., 1982, 1983], or give rise to various extravillous trophoblast populations. A full transition to mesenchymal phenotype is observed in only one trophoblast subpopulation: cytotrophoblast of the basal decidua that has infiltrated maternal mesenchymally-derived uterine stroma and arterial walls [Damsky et al., 1993; Denker, 1993]; this is not the case in other differentiated cytotrophoblast populations that remain more directly in association with the placenta (such as the cytotrophoblast multilayer of the chorion laeve).

Ontogeny of Trophoblast EMT

In the earliest post-implantation phases, many textbooks depict the outermost layer of the embryo as one continuous syncytium. However, it probably comprises a multiplicity of syncytial cells together with closely associated cytotrophoblast [Hertig et al., 1956]. In any case, from day 13 post-ovulation and probably earlier, both cyto- and syncytiotrophoblast interact directly with the uterine stroma [Enders, 1991]. This complex mixture of trophoblast actively colonises and remodels the uterine environment. After day 21 post-ovulation a more regular morphology is apparent in which the fetal mesenchyme, now vascularised, is covered by a monolayer of cytotrophoblast on the outside of which is a continuous syncytium (fig. 1c, 2). Maternal blood eventually circulates in the intervillous spaces from which transport of nutrients to the fetus occurs across the villous syncytium. However, there is evidence to suggest that this vascular circuit may not be not fully functional until as late as the 11th–13th week of pregnancy [Schaaps and Hustin, 1988].

In the earliest stages of placental development, trophoblast proliferation and invasion is very rapid [Enders et al., 1996]. There are large numbers of interstitially and intravascularly migrating cells, both mono- and multinucleate. There is an irregular boundary of contact between the trophoblastic shell and maternal stroma (fig. 1a, b). Eventually, with the expansion of the intervillous spaces, specific sites of placental attachment become apparent, known as 'anchoring villi' (fig. 1c, 2). These are characterised by cy-

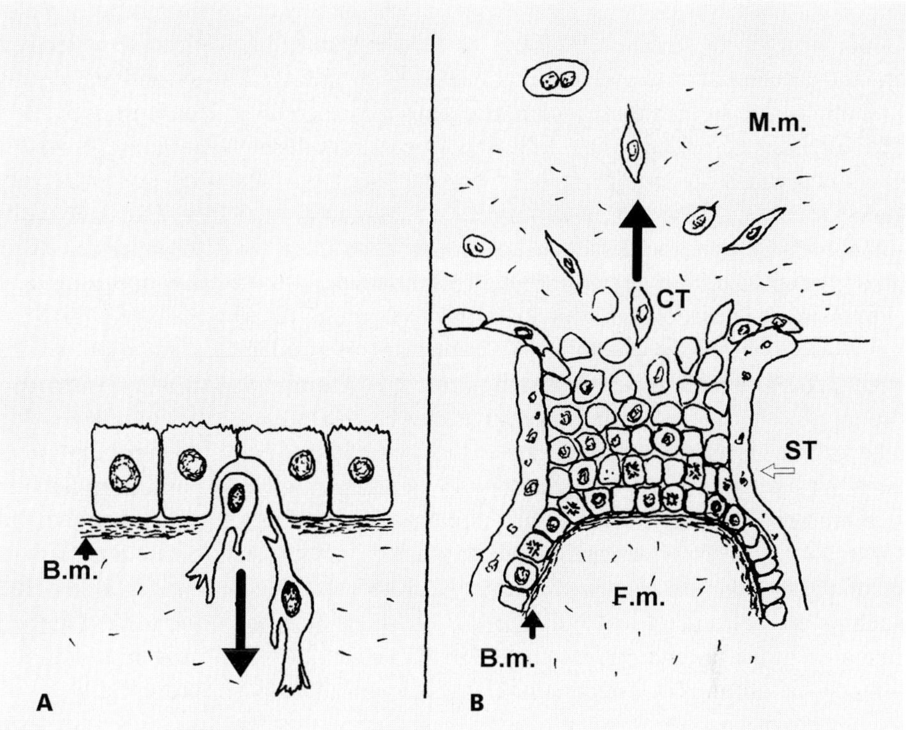

Fig. 3. Schematic representations of EMTs in which (**A**) epithelial cells degrade their own basement membrane (B.m.) to infiltrate underlying stroma [redrawn after Hay, 1991; see Hay, 1995]; and (**B**) cytotrophoblast (CT) forms multilayers at sites of anchorage, breaches the syncytiotrophoblast (ST) and detaches to infiltrate maternal mesenchymal stroma (M.m.). In **B** neither the villous basement membrane (B.m.) nor fetal (extraembryonic) mesenchyme (F.m.) is breached. Note that, for purposes of comparison, the orientation in **B** is opposite to that used in figure 2.

totrophoblast columns from which single cytotrophoblast cells continue to detach and invade the maternal decidua (fig. 1d, 2) up until about the 18th week [Pijnenborg et al., 1980], by which time the most advanced cells have penetrated as far as the innermost third of the myometrium. It is this subpopulation of the extravillous trophoblast (also referred to as interstitial, intermediate [Kurman et al., 1984], or invading trophoblast) that has undergone EMT. Its major function appears to be the remodelling of maternal arteries [Pijnenborg et al., 1980]. This involves wholesale changes to the normal mural architecture, with the loss of smooth muscle and associated elastic and collagenous extracellular matrix, these being replaced by a fibrinoid matrix, allowing an increased and stabilised flow of blood to the placenta. Placental giant cells are seen deep in the placental bed and myometrium; these appear to represent a terminally differentiated interstitial-type trophoblast.

Morphological Characteristics of Trophoblast EMT

Compared to other EMTs in the embryo, where epithelial cells in the course of transition degrade their own basement membrane to infiltrate underlying stroma (fig. 3A), the direction is the opposite in the case of trophoblast (fig. 3B). Cytotrophoblast progenitor cells that reside upon the villous basement membrane, break through the overlying syncytium specifically at sites of anchorage to the uterine wall, dividing rapidly [Vićovac et al., 1995a, b] to form columns of cytotrophoblast as a result of local proliferation. Since it is only the most proximal cytotrophoblast layers that proliferate, the daughter cells are displaced further away from the villous basement membrane towards the maternal decidual stroma. In the course of these events the cells become less polarised, larger and round, and in the medial and distal columns the cytotrophoblast becomes enmeshed in a new extracellular matrix. At the interface with maternal stroma, and in deeper locations, both rounded and elongated cytotrophoblast can be seen. There are elements reminiscent of EMT seen in vitro using three-dimensional matrices [Hay, 1995].

Mesenchymal Host Tissue

It has often been said that the penetrative nature of human placentation mimics that seen in highly invasive tumors, the main difference being that there is a strict control in the extent of trophoblast invasion. Thus, the uterine milieu, while nourishing and supporting the conceptus (especially in the first trimester before the haemochorial placenta is fully functional) must also exert firm control of invasion [Strickland and Richards, 1992]. Paracrine regulatory signals from decidualised endometrial stroma are likely to be important, and indeed it is clear from knock-out experiments in mice that maternal rescue of an abnormal fetal phenotype can occur (e.g. TGFβ_1) [Letterio et al., 1994].

Initially, the human embryo implants into mid-secretory endometrial stroma. Decidual differentiation occurs during the first few days post-implantation. The decidua is a complex tissue [Starkey, 1993] comprised of glandular and surface epithelium, the decidualised stromal cells of fibroblastic origin, and numerous immune cell populations. Decidualisation is associated with an influx of immune cells, morphological and biochemical differentiation of resident stromal cells and the onset of specific protein and hormone production by decidual cell subpopulations. The differentiated stromal cells produce new extracellular matrix components including laminin, collagen type IV and heparan sulfate proteoglycan [Aplin et al., 1988; Aplin and Jones, 1989; Church et al., 1996b] which become organised into a pericellular basal lamina. They also make secretory products such as prolactin (PRL) [Riddick et al., 1978] and insulin-like growth factor binding protein 1 (IGF-BP1) [Bell, 1989]. Among the immune cell population, large granular lymphocytes and macrophages are present, as well as a smaller number of T cells [Bulmer and Sunderland, 1984; Bulmer and Johnson, 1985; Starkey et al., 1988]. Their function is unknown but it has become apparent that they produce cytokines [Tabibzadeh, 1994; Mitchell et al., 1993]. Thus, in first trimester, decidual immune cells release CSF-1, TNFα, IFNγ, TGF-β_1, LIF and GM-CSF [Vince et al., 1992; Jokhi et al., 1994a]. Cytokines have increasingly been implicated in trophoblast differentiation (see below), and their occurrence at the feto-maternal interface is of significance.

Molecular Criteria of the Epithelial-Mesenchymal Differentiation Pathway in Trophoblast: In vivo Descriptive Studies

Cell Adhesion Molecules

As cells develop from polarised stem cytotrophoblast attached to the villous basement membrane into columns and then to interstitially migrating cytotrophoblast, radical alterations occur in cell-cell and cell-matrix interactions. Some of the molecules involved in these interactions are well recognised, while understanding of the underlying regulatory mechanisms is lacking. Within the normal villus, the interface between the apical cytotrophoblastic surface and the basal membrane of the syncytium is characterised by prominent desmosomes [Beham et al., 1988], and the calcium-dependent homotypic adhesion molecule E-cadherin [Fisher et al., 1989]. E-cadherin seems to play an important role in early (mouse) development and implantation since transgenic mice lacking the gene fail to form trophectoderm and do not implant [Larue et al., 1994]. The villous cytotrophoblast layer also expresses the α_6 and β_4 integrin subunits [Korhonen et al., 1991; Damsky et al., 1992; Aplin, 1993]. During the epithelial-mesenchymal transition of trophoblast, E-cadherin is downregulated [Babawale et al., 1996] as has been observed in other EMTs [Burdsal et al., 1993]. Strong upregulation of expression of a polysialylated form of NCAM has been observed in the endovascular subpopulation of extravillous trophoblast [Burrows et al., 1994]. Cytotrophoblast cells that leave the villous basement membrane to form cell columns also undergo a striking downregulation of α_6 and β_4 integrins [Aplin, 1993] accompanied by the upregulation of α_5 and β_1 integrin subunits [Fisher et al., 1989; Damsky et al., 1992]. Integrin $\alpha_5\beta_1$ has been identified as a key marker of the mesenchymal phenotype in other EMTs [Hay, 1995]. Within the uterine wall, the mesenchymally transformed interstitial cytotrophoblast switch on a new integrin subunit – α_1 [Zhou et al., 1993], presumably giving rise to the heterodimer $\alpha_1\beta_1$ which acts as a receptor for laminins and/or collagens. In pre-eclampsia, intravascular cytotrophoblast invasion has been reported to be abnormally shallow and the transformation of maternal arteries incomplete [Pijnenborg, 1994]. It has been reported that the downregulation of integrin $\alpha_6\beta_4$ and upregulation of $\alpha_1\beta_1$ fails in placental bed cytotrophoblast in this condition [Zhou et al., 1993].

Extracellular Matrix (ECM)

Clearly, the availability of the respective ECM ligands for integrins at the site of invasion is a consideration in the interpretation of these changes. Mesenchymal and epithelial states are usually associated with production of distinct repertoires of ECM molecules; an additional complication arises in the case of trophoblast as a result of the fact that during the EMT, cells interact with an environment with progressively more maternal contribution. Trophoblast at the villus attach via the basal surface to a basement membrane of (presumably) extraembryonic origin that contains collagen IV and heparan sulphate proteoglycan. In addition, there is an extensive repertoire of laminin isoforms: subunits α_1, α_2, α_4, β_1, β_2 and γ_1 are all detectable [Damsky et al., 1992; Ohno et al., 1986; Leivo et al., 1989; Engvall et al., 1990; Church et al., 1996a], suggesting that laminins 1, 2, 3 and 4 [Burgeson et al., 1994] and others may be present. The underlying extraembryonic mesenchyme contains both interstitial-type and basement membrane-type matrix components. However, its composition is not homogeneous, tenascin is expressed at sites of cytotrophoblast cell mitosis (such as beneath columns) and under degenerative syncytiotrophoblast [Castellucci et al., 1991]. It is not clear whether this reflects cause or effect; villous core (i.e. extraembryonic) mesenchymal or basement membrane specialisation at sites of column formation may play a role in the formation of anchoring villi; alternatively, it may be a reaction to cytotrophoblast proliferation and differentiation [King and Blankenship, 1994]. However, tenascin's anti-adhesive properties might allow trophoblast to increase migratory activity along the villus or into columns [Aufderheide and Ekblom, 1988].

As the cells proliferate at anchorage sites to form columns, they lose their polarised epithelial characteristic, and, beyond the first few cell layers, an unusual ECM appears that is at least partly of cytotrophoblastic origin. This contains no fibrillar collagen, as assayed by electron microscopy [Enders, 1968], but is rich in basement membrane components including collagen IV, heparan sulphate proteoglycan and laminin [Damsky et al., 1992; Church et al., 1996a]. There are rich deposits of 'oncofetal' fibronectin [Feinberg et al., 1991]. This carries a glycopeptide epitope associated with fibronectin produced by cytotrophoblast in vitro, but it is scarce in maternal plasma fibronectin and in decidual fibronectin; thus it is likely that the columnar cytotrophoblast is responsible for producing it.

Once the cells leave the columns and enter the maternal interstitial environment, less is known about their production of extracellular matrix. At sites where columns disperse to form migratory cells, the trophoblastic extracellular matrix is intimately associated with the extracellular matrix of the maternal decidua, and with another extracellular deposit known as 'fibrinoid' which accumulates in places to form a discrete layer known as Nitabuch's stria. Fibrinoid is ultrastructurally distinct from fibrin [Frank et al., 1994]. It is also found in association with some floating villi, possibly functioning as a repair matrix at sites of damage to the trophoblast epithelium [Nelson et al., 1990; Farmer and Nelson, 1992]. It seems possible that, in these specialised locations, fibrinoid acts in place of interstitial-type collagens as an ECM with roles in both mechanical integrity and cellular interactions. The intra-arterially migrating cytotrophoblast effects extensive degradation of the musculoelastic matrix of the vessel walls which is replaced by fibrinoid. Intravascular trophoblast of the human [Fernandez et al., 1990] and the macaque [Blankenship et al., 1993] has also been shown to produce a laminin- and collagen IV-rich capsule. Vascular transformation allows a greatly increased flow of blood to the intervillous space and also renders the blood supply to the placenta independent of vasoconstrictors. This vascular conversion, effected by a combination of intra-arterially and interstitially migrating cells, is likely to be the major function of the extravillous trophoblast population.

The ECM of maternal decidua contains collagens I, III, V and fibronectin, while individual decidual cells elaborate a pericellular basal lamina of collagen type IV, laminins 2 and 4 and heparan sulphate proteoglycan [Wewer et al., 1985; Aplin et al., 1988; Aplin and Jones, 1989; Aplin et al., 1995; Church et al., 1996b].

Proteases and Inhibitors

In blastocyst implantation and trophoblast invasion there is a role for proteolysis at several steps: at the time of penetration of the epithelial basement membrane after embryo apposition, later when individual trophoblast cells invade decidual stroma and during vessel wall remodelling by intravascular trophoblast. Localised proteolysis appears to be achieved by a combination of autocrine and paracrine regulation at the enzyme, substrate and inhibitor levels. There is much evidence to demonstrate that trophoblast produces different proteases capable of local degradation and/or invasion of extracellular matrix including plasminogen activator, interstitial collagenase (MMP1) and 72-kD (MMP3) and 92-kD (MMP9) type IV collagenases [Fisher et al., 1985, 1989; Yagel et al., 1988; Lala and Graham, 1990; Moll and Lane, 1990; Autio-Harmainen et al., 1992; Polette et al., 1994]. 72-kD type IV collagenase has been localised to intra-arterial cytotrophoblast during invasion of utero-placental arteries in the macaque [Blankenship and King, 1994]. Another type of proteinase, urokinase-type plasminogen activator (uPA) is made by cytotrophoblast [Queenan et al., 1987]. In situ localisation has not revealed significant qualitative differences between trophoblast subpopulations. Interstitial collagenase has been detected in all trophoblast [Moll and Lane, 1990] and 72-kD type IV collagenase localises in both villous and extravillous cytotrophoblast, as well as in syncytiotrophoblast [Autio-Harmainen et al., 1992]. However, the cytotrophoblast of the columns express the 72- and 92-kD enzymes particularly strongly [Polette et al., 1994]. In addition, proteolytic activity has been found in decidual tissue [Autio-Harmainen et al., 1992; Schatz et al., 1994]. Stromelysin 3 has an interesting distribution, being found in both villous syncytiotrophoblast and the interstitially invading cytotrophoblast population, but not in the columns [Maquoi et al., 1996]. The actions of these enzymes are regulated by inhibitors within the local environment including the tissue inhibitors of metalloproteinase (TIMP1 and TIMP2) produced by both trophoblast and decidua [Lala and Graham, 1990; Polette et al., 1994] or plasminogen activator inhibitor (PAI-1) produced by the invasive extravillous trophoblast itself [Feinberg et al., 1989]. Plasminogen activator is required to produce plasmin, and this in turn activates latent collagenases. Trophoblast has been shown to express two types of inhibitors of plasminogen activation in vivo and in vitro, with PAI-1 predominantly in invading trophoblast and PAI-2 in syncytiotrophoblast [Feinberg et al., 1989]. Mice that fail to express PAI-1 are normally fertile, with no abnormalities of pregnancy [Carmeliet et al., 1994]; however, mouse trophoblast does not infiltrate the maternal decidual stroma, and indeed the term EMT is not really appropriate to describe its behaviour. Thus it remains unknown whether the plasminogen/PA system is required for human placentation.

Trophoblast has also been shown to produce α_1-antichymotrypsin and α_1-antitrypsin, closely related glycoprotein protease inhibitors, present in plasma and other extracellular fluids, that neutralise proteases released by leucocytes in response to trauma and inflammatory stimuli [Bergman et al., 1993].

Other Cell Markers: Epithelial vs. Mesenchymal

Along the differentiation pathway that leads to the interstitial/mesenchymal trophoblast, certain of the characteristics of villous cytotrophoblast are lost; others are retained while specific new ones are acquired.

Cytokeratin. In mouse, cytokeratin is first detected at the compacted morula stage in cells destined to become trophoblast [Lehtonen et al., 1983] and its expression is maintained throughout pregnancy. Similarly, cytokeratins are expressed in all human trophoblast populations throughout gestation. While cells involved in certain other EMTs cease to express cytokeratins and switch on vimentin instead [Hay, 1995], infiltrating trophoblast continues to express a repertoire of cytokeratins [Mühlhauser et al., 1995] and there are no well-documented reports of vimentin expression. Thus, cytokeratin is a marker that facilitates discrimination between trophoblast and other cell types during infiltration of decidua.

HLA-G. The syncytiotrophoblast and stem cytotrophoblast of first trimester placenta do not express MHC class I antigens. It is thought that this avoids rejection by the maternal immune system. However, cytotrophoblasts of the distal columns, as well as the interstitial and endovascular extravillous cytotrophoblasts, carry non-polymorphic framework determinants of HLA-A, -B, and -C, but not the polymorphic class I determinants [Redman et al., 1984; Loke and Butterworth, 1987; Ellis et al., 1990; Lala et al., 1990]. Further characterisation has led to the recognition of HLA-G, an apparently monomorphic form of class I that has also been detected in adult peripheral blood lymphocytes [Kirszenbaum et al., 1994]. HLA-G was identified in extravillous trophoblast populations including the cytotrophoblast cell columns, interstitial trophoblast, endovascular trophoblast, and placental bed giant cells [Chumbley et al., 1993; McMaster et al., 1995].

NDOG-5. This antigen, though not biochemically characterised is highly specific to extravillous trophoblast [Shorter et al., 1993]. It is common to the distal columns, shell and invasive cytotrophoblast.

Control of the Trophoblast Epithelial-Mesenchymal Transition

Experimental Models

3D Cultures and Cocultures. The study of early morphogenetic events in human placentation is beset by ethical and practical problems. New insights or hypotheses depend on the use of in vitro models. A coculture technique has been described that allows adhesion and further interactions between villous tissue and decidua to be studied [Aplin et al., 1993; Vićovac et al., 1993, 1995a]. Explanted first-trimester villous tissue fragments readily attach to decidual stroma (tissue selected from non-placental bed locations to avoid endogenous trophoblast). The outgrowths produced are of trophoblast origin. Over the following days de novo differentiation occurs into cytotrophoblast columns and the invasive lineage. The cells express the extravillous cell markers NDOG5 and HLA-G, show accumulations of ECM components typical of columns in vivo, and undergo a stepwise progression of altered integrin expression including the loss of $\alpha_6\beta_4$ and the acquisition of $\alpha_5\beta_1$, exactly as observed for cytotrophoblast of anchoring villi in vivo. Subsequently, the cells invade the adjacent decidua (fig. 1e). However, not all sites of column formation are associated with a mesenchymal-type invasive cell population. Similar observations are made when decidual tissue is substituted with peritoneum [Vićovac et al., 1993]. These observations support a model in which the formation of columns precedes acquisition of the fully mesenchymal phenotype. It appears likely that paracrine signals from decidua may play a role in triggering migratory behaviour. Thus, the tips of late first-trimester floating villi retain the ability to form new anchoring sites and produce infiltrating cells if placed in apposition to decidua (fig. 1e). This could be interpreted as a recapitulation of earlier developmental events, but it may also be that this capacity continues to be relevant to placental development up to the end of the first trimester.

In general, systems that provide a source of 3D support matrix (Matrigel, collagen or decidual tissue) also allow the outgrowth of extravillous cytotrophoblast (see below). When first-trimester placental villi are cultured as explants on Matrigel or collagen type I gel, they produce cells (fig. 1f) that express extravillous cytotrophoblast markers [Genbačev et al., 1992; Vićovac et al., 1995b]. Thus it appears that the attachment of a villus tip to ECM is in itself capable of triggering the formation of columns.

Trophoblast Primary Cell Culture and Strains. First-trimester placental villi explanted onto tissue culture plastic or carpets of extracellular matrix produce cytotrophoblast of extravillous lineage, though at variable rate and purity [Fisher et al., 1985; Yagel et al., 1988, 1989]. A proportion of cultures obtained in this way from first trimester tissue are pure trophoblast while there are some in which there is fibroblast contamination [Irving et al., 1995]. Cultures of placental villi on plastic and in anaerobic conditions give rise to proliferating preparations of cytotrophoblast [Yagel et al., 1989]. Such cell strains are reported to retain some of the characteristics of extravillous trophoblast for 12–14 passages, and, following transfection with SV40 large T antigen, for longer than 32 passages [Graham et al., 1993].

More typically, isolation protocols include trypsin digestion of villous tissue [Kliman et al., 1986] or treatment with collagenase and hyaluronidase [Fisher et al., 1989]. It is necessary to follow this with a further purification step using, for example, density gradient centrifugation or antibody-coated beads to remove contaminating cell populations (such as syncytial remnants, erythrocytes, fibroblasts and bone marrow-derived cells). Such preparations have been immortalised by transformation with temperature-sensitive Simian virus 40 [Chou, 1978; Logan et al., 1992].

A key question about all trophoblast preparations concerns the origin of the cells obtained; depending on the precise combination of methodology and stage of pregnancy, they could be either villous stem cytotrophoblast or cells from cytotrophoblastic shell, columns and islands which are comparatively abundant in first-trimester material. On the other hand, first-trimester villous stem cytotrophoblast is also more abundant per unit area of villus than at term [Benirschke and Kaufmann, 1990]. First-trimester cultures contain a non-proliferative mononuclear cell population resembling extravillous trophoblast. Thus, for example, Burrows et al. [1993] and Irving and Lala [1995] showed by flow cytometry that their first-trimester cells expressed integrin subunits α_5 and β_1, characteristic of the medial and distal columnar and mesenchymal phenotypes.

Trophoblast subpopulations can be isolated from first-trimester placenta by brief trypsinisation followed by 3–4 days of culture. By this time the adherent cell population has differentiated into syncytia. However, small cell aggregates are found in suspension; collection and further culture of these leads to an adherent mononuclear extravillous cell type with an elongated bipolar morphology [Aboagye-Mathiesen et al., 1994].

A frequently used marker of the differentiation state of cultured trophoblast is hCG, which in vivo is largely specific to the syncytial pathway [Morrish et al., 1987; Kliman et al., 1986]. However, cytotrophoblast isolated from villous outgrowths and cultured for several passages, while showing characteristics of the invasive trophoblast, has also been shown to produce hCG in vitro [Yagel et al., 1989], and the same is true after transformation with SV40 [Graham et al., 1993]. The presence of hCG may reflect a slow rate of differentiation into syncytiotrophoblast; most cultures do contain a proportion of these cells. It might also reflect a drift into a mixed phenotype that does not represent any one in vivo phenotype with high fidelity [Kliman and Feinberg, 1992].

Characterisation of the newly isolated cells is also complicated by the likelihood that they may change rapidly on release from the tissue. Isolated term cytotrophoblast spreading on tissue culture plastic or other rigid substrata tend rapidly to follow the syncytial differentiation pathway [Kliman et al., 1986; Morrish et al., 1987; Douglas and King, 1989; Yui et al., 1994], or exhibit a combination of traits characteristic of syncytial and invasive cell types [Kliman and Feinberg, 1992]. When the cells are seeded onto gel matrices – matrigel, collagen or fibrin – the tendency to form syncytia is reduced. These data suggest that the physicochemical properties of the substratum influence morphological differentiation [Hohn et al., 1993; Farmer and Nelson, 1992]. Cultured cytotrophoblast has been used to study invasive characteristics, as a way of modelling trophoblast invasion. Cells seeded onto a Matrigel- (or collagen-) covered filter first aggregate, then penetrate the gel and through pores to the far side of the filter. A decline occurs in invasiveness of cells isolated from placentas closer to term [Fisher et al., 1989; Librach et al., 1991]. The relevance of integrin expression for invasive behaviour was confirmed in this assay [Damsky et al., 1994] by using function-blocking antibodies. It was observed that addition of fibronectin inhibited migration of the cells through the gel and across the filter. Addition of function-blocking antibodies to integrin subunits α_5 or β_1 enhanced trans-filter migration. This led to the conclusion that interaction of cytotrophoblast with fibronectin, mediated by $\alpha_5\beta_1$ integrin, inhibits migration. In contrast, interaction of integrin $\alpha_1\beta_1$ with its ligands (collagen IV and laminin) is required for invasion to occur. In vivo, integrin $\alpha_5\beta_1$ is expressed on cells of the medial and distal columns as well as the migrating cytotrophoblast; integrin $\alpha_1\beta_1$ is present only on the interstitially invading (mesenchymal-type) trophoblast after they leave the distal tips of the columns [Korhonen et al., 1991; Damsky et al., 1992; Aplin et al., 1993b]. These data can be interpreted as indicating that there are at least two functionally distinct stages in the development of the extravillous lineage: the columns are true anchoring structures in which the cell-fibronectin interaction plays a stabilising and linking role, while the truly invasive cells represent a further differentiative transition enabling interactions with integrin $\alpha_1\beta_1$ to mediate migration through a laminin and collagen IV-rich environment provided mainly by decidual cells.

Proteolysis is also required for trophoblast penetration through artificial barriers, either cell-free amniotic membrane [Yagel et al., 1988], or matrix-coated filters [Fisher et al., 1989; Librach et al., 1991]. First-trimester trophoblast produces the 92-kD type IV collagen-degrading metalloproteinase (MMP9), and this has been shown to be required for invasiveness in vitro [Librach et al., 1991]. Its synthesis and activation peak in first trimester and correlate well with the period of invasive behaviour of trophoblast in vivo. Metalloproteinase inhibitors and a function-perturbing antibody specific for the 92-kD metalloproteinase completely inhibit cytotrophoblast invasion of Matrigel gels. Inhibitors of the plasminogen activator system reduce but do not completely abolish invasion in the Matrigel penetration assay, also consistent with the involvement of collagenases in the process.

The composition of the ECM can affect the profile of proteolytic enzymes produced by first-trimester cytotrophoblast [Emonard et al., 1990]. The cells show reduced activity when cultured on plastic compared to collagen type I gel or Matrigel, but production of type IV collagen degrading activity is stimulated by addition of soluble laminin into the culture medium. Cells cultured on 3D gels produce in general much higher amounts of enzymes and no difference is observed between Matrigel and collagen type I. Zymographic analysis of matrix-degrading metalloproteinases produced by first-trimester trophoblast cultured on Matrigel, collagen type I and agarose [Bischof et al., 1991] identified 4 enzymes secreted into the medium, one of which, a 59-kD gelatinase, was specific to cells cultured on Matrigel. However, results based on models that utilise Matrigel have to be treated with caution, since it contains endogenous proteolytic activity, including the 72- and 92-kD collagenases [Mackay et al., 1993] as well as growth factors which could also influence cell phenotype [Vukicevic et al., 1992].

Growth Factors and Cytokines

A large range of growth factors and cytokines has been detected in the different cell populations at the maternal-fetal interface. Here we restrict discussion to components that have been shown to modulate the behaviour of trophoblast in the extravillous lineages that result from EMT.

EGF Family. EGF has been shown to bind to the apical syncytial surface of first-trimester placental villi from where transcytosis occurs to the underlying cytotrophoblast [Arnholdt et al., 1991]. The EGF receptor (c-erb-B1) is detectable in the villous cytotrophoblast and syncytium, as well as in the proximal regions of cytotrophoblast columns [Bulmer et al., 1989; Filla et al., 1993; Mühlhauser et al., 1993]. The related tyrosine kinase receptor c-erb-B2 is expressed in a complementary pattern on the medial and distal cells of the columns [Mühlhauser et al., 1993; Jokhi et al., 1994b]. In situ, it is only villous stem cytotrophoblast and several most proximal cell layers of the cell columns that proliferate [Castellucci et al., 1991]. Furthermore, it is widely observed that isolated cytotrophoblast cells do not readily proliferate in vitro. However, there is some evidence that EGF can induce trophoblast proliferation in vitro, at least in explant cultures from very early (4–5 week) placenta [Maruo et al., 1992]. In contrast, cultures from later stages of gestation respond to EGF by increasing the production of hCG and PL [Morrish et al., 1987; Maruo et al., 1992]. EGF also significantly enhances the ability of cultured first-trimester cytotrophoblasts to cross a Matrigel barrier, apparently by increasing cell motility, though this effect is relatively short lived (12–24 h in culture) and diminishes with gestational age [Bass et al., 1994]. EGF therefore appears to stimulate differentiation in either the syncytial or the extravillous lineages in vitro, depending on the characteristics of the target cells. TGFα, which acts through the EGF receptor, has been reported to stimulate proliferation of second-passage first-trimester cytotrophoblast [Lysiak et al., 1993].

It is not clear what the physiologically relevant ligand(s) for the EGF receptor in trophoblast may be. Bass et al. [1994] were unable to detect EGF transcripts by RT-PCR in trophoblast. However, there is evidence for the presence of both EGF and TGFα in decidual cells [Haining et al., 1991; Lysiak et al., 1993] indicating the possibility of a paracrine interaction. TGFα has also been detected immunochemically in villous and extravillous trophoblast at all gestational ages [Hofmann et al., 1993; Lysiak et al., 1993]. There is evidence in rodent implantation sites for the expression of other ligands that act through the EGF receptor e.g. HB-EGF [Zhang et al., 1994].

TGFβ. Antibodies that bind $TGF\beta_2$ or both $TGF\beta_1$ and $TGF\beta_2$ have been shown to label first-trimester decidua (both the cells and ECM), as well as villous mesenchyme, syncytiotrophoblast and the cytotrophoblast shell [Dungy et al., 1991; Graham et al., 1992; Chegini et al., 1994; Selick et al., 1994]. Cultured first trimester trophoblast secretes a latent form of $TGF\beta_1$ [Matsuzaki et al., 1992]. There are several TGFβ receptors in placental tissue, including receptors type I, type II, betaglycan and endoglin [Mitchell et al., 1991, 1994; St. Jacques et al., 1994]. Endoglin is restricted to the apical syncytial plasma membrane. When normal or transformed cytotrophoblast with extended life span have been derived, their rate of division is decreased in the presence of TGFβ [Graham et al., 1993]. Similarly, rat trophoblast cell lines are inhibited from proliferation by TGFβ [Hamlin and Soares, 1995]. TGFβ seems to be a potent regulator of trophoblast differentiation. It has been shown to promote the deposition of oncofetal fibronectin by term trophoblast in vitro [Feinberg et al., 1994]. TGFβ upregulates gelatinase A mRNA in normal trophoblast and JAR choriocarcinoma cells; it also significantly upregulates TIMP1 and TIMP2 mRNA in normal trophoblast. Its ability to inhibit invasiveness in vitro is probably mediated by TIMPs since TGFβ induces an increase in mRNAs for TIMP1 and TIMP2 [Graham and Lala, 1991]. However, other investigators could not demonstrate that it inhibits invasiveness in vitro [Bass et al., 1994]. PA secretion in vitro is also reduced in the presence of TGFβ. There is conflicting evidence regarding a possible influence on the syncytial differentiation pathway with some authors reporting inhibition in vitro [Morrish et al., 1991], while others found stimulation of syncytium formation [Graham et al., 1992].

IL-1. IL-1β has been immunolocalised in decidual cells and trophoblast in situ [Kauma et al., 1990]. It is produced by freshly isolated first-trimester cytotrophoblast cells during the first 24 h in culture and it upregulates metalloproteinase activity and invasive activity of these cells in vitro, thus potentially acting in an autocrine loop [Librach et al., 1994]. It also stimulates the release of hCG in culture [Masuhiro et al., 1992]. IL-1α has been localised in villous trophoblast [Hu et al., 1992]. Interest in the role of IL-1 in human has been increased by the finding that IL-1 receptor antagonist blocks implantation in mice [Simon et al., 1994].

IGFs. IGF II mRNA is strongly expressed in first-trimester extravillous trophoblast, and more weakly in villous cytotrophoblast and decidual cells. IGF I is associated with the villous syncytium. The receptor IGF IR is present on trophoblast populations with the exception of the syncytium and on decidual and fetal mesenchymal cells. The IGF binding proteins IGFBP1 and IGFBP3 are expressed in decidual cells and extravillous cytotrophoblast, respectively [Glaser et al., 1992; Ekström et al., 1994]. The presence in the IGFBP1 of an active RGD motif that interacts with integrin $\alpha_5\beta_1$ to promote cell adhesion may also be relevant to the behaviour of migratory cytotrophoblast in which this receptor is upregulated [Irving and Lala, 1995]. IGF II has been reported to stimulate cell cycle entry or progression in serum-starved trophoblast cultures [Ohlsson et al., 1989] and to enhance cytotrophoblast invasion [Lysiak et al., 1993; Irving and Lala, 1995]. This enhancement is further increased by the presence of IGFBP1 [Irving and Lala, 1995]. These data suggest the existence of complex autocrine or juxtacrine regulatory loops.

CSF-1, GM-CSF. Another cytokine potentially relevant for trophoblast differentiation is colony-stimulating factor-1 (CSF-1) which is localised in the villous syncytium [Haynes et al., 1993]. Its receptor is absent from villous cytotrophoblast, but it is strongly expressed in cells of the columns and weak or absent in trophoblast present in deeper regions of decidua [Jokhi et al., 1993]. In culture, CSF-1 and GM-CSF stimulate trophoblast aggregation and differentiation along the syncytial pathway [Garcia-Lloret et al., 1994]. GM-CSF binds to first-trimester trophoblast cultures causing increased DNA synthesis [Loke et al., 1992]. Further, the GM-CSF receptor has been identified on first-trimester trophoblast [Jokhi et al., 1994a]. GM-CSF is produced both by cultured trophoblast and decidual large granulated lymphocytes [Jokhi et al., 1994c].

LIF. Mice lacking leukaemia inhibitory factor fail to support blastocyst implantation [Stewart et al., 1992]. However, the –/– blastocytes implant and develop normally in wild-type recipient females. This is clear evidence for an obligate maternally derived signal in implantation, though it is not yet clear to what it is directed. LIF peptide and mRNA have been detected in glandular epithelium of both mouse and human and are reported to be higher in the peri-implantation than in the proliferative phase [Charnock-Jones et al., 1994b] and to persist in pregnancy [Kojima et al., 1994].

VEGF. VEGF increases thymidine incorporation by BeWo choriocarcinoma cells; its receptor flt is found on cells of the extravillous lineage in situ suggesting that it may be involved in stimulating trophoblast proliferation in vivo [Charnock-Jones et al., 1994a].

Scatter Factor. Scatter factor, also known as hepatocyte growth factor (HGF) is expressed in the villous syncytium and extravillous cytotrophoblast of human placenta [Wolf et al., 1991]. Placental deficiencies including reduced labyrinthine trophoblast are evident in mice that lack HGF [Schmidt et al., 1995; Uehara et al., 1995].

TNFα. TNFα is found in the villous syncytium [Haynes et al., 1993] and in cytotrophoblast and decidual cells [Yang et al., 1993]. It may be involved in autocrine/paracrine regulation of trophoblast growth [Yang et al., 1993]. Bass et al. [1994] observed no effect on invasiveness of first-trimester cytotrophoblast in vitro.

PDGF. PDGFA mRNA is detected in first-trimester cytotrophoblasts of the columns [Holmgren et al., 1991], and PDGF a and b receptors are found on cytotrophoblast [Holmgren et al., 1992]. Addition of PDGF-AA or -BB to cultured cytotrophoblast increased DNA synthesis [Holmgren et al., 1992] but not penetration of Matrigel [Bass et al., 1994].

hCG. Intra-trophoblastic juxtacrine activity might be suggested by the finding that complexation of hCG in culture medium with an antibody resulted in a decrease in DNA synthesis. Furthermore, plasminogen activation and collagenolytic activity decreased in cultured cytotrophoblast in the presence of high levels of hCG, which is produced mainly by syncytiotrophoblast in vivo [Yagel et al., 1989, 1993]. Gap junctional communication during differentiation in the syncytial lineage is also increased by hCG [Cronier et al., 1994].

Conclusion: Current Understanding of the Epithelial-Mesenchymal Transition in Trophoblast

Investigations to date have established that a stem cytotrophoblast population at the villous basement membrane can differentiate to follow either the syncytial (epithelial) or extravillous (mesenchymal) lineage. Considerable advances have been made in establishing some of the characteristics of these derivative cell populations. A further population of non-invasive cytotrophoblast arises in the chorion laeve; this represents another differentiation pathway but has not been considered here since it does not appear until late in the first trimester [Aplin, 1991]. Advances have also been made in defining the presence of cytokines and growth factors and their receptors in the diverse interacting cell populations in the placental bed. However, essentially nothing is yet known of the molecular mechanisms controlling the choice of differentiation pathway made by the trophoblastic stem cells.

Whether all villous cytotrophoblast is equipotent is not established. However, there is little to suggest heterogeneity in the villous cytotrophoblast population. Since cytotrophoblast isolated from term placenta aggregate and fuse within 96 h in vitro, it may be that the interaction with villous basement membrane or stroma has the effect of inhibiting differentiation in the absence of an overriding stimulus. This might be coupled to cell shape, since substrates that disallow cell spreading seem to suppress fusion. There is evidence to support the view that column formation occurs preferentially from the smallest terminal (mesenchymal) villi (fig. 1f). It is therefore interesting that these villi contain a compositionally distinct mesenchyme (highly cellular, tensacin-positive and with a thickened basement membrane). Thus a signal from villous extraembryonic mesenchyme might be required to permit columm formation by trophoblast. Given the spatial relationships between anchoring villi and decidua, it is also likely that signals from the opposite direction – from maternal decidua – can stimulate the rapid formation of columns at sites of villous anchorage. Data from a model (fig. 1e) in which villous tissue is maintained in vitro in contact with decidua suggest further that column formation is a prerequisite of the further differentiation of trophoblast into mesenchymal-type cells that infiltrate the decidual interstitium [Vićovac et al., 1995a]. Experiments with cultured cytotrophoblast coupled with in vivo localisation data have identified several candidate molecules that may play a role in the paracrine interactions that both stimulate the mesenchymal transformation of trophoblast and limit its invasion to the decidua and inner myometrium. However, uncertainty as to the origins and differentiation state of the cells isolated, the loss of villous and columnar organisation in vitro and the likelihood that rapid changes of phenotype occur, mean that these data must be treated with caution. Most of all, there is a need for in vitro systems in which uncommitted trophoblast can be manipulated to alter developmental fate. There is strong evidence for an inherent programme of trophoblast differentiation in which invasiveness diminishes with gestational age [Aplin, 1991], and the effects of the maternal environment must therefore be considered in the context of progressive alteration in the invasive cell phenotype. Improved models of the human trophoblast-decidua interaction [Vićovac et al., 1995b] should allow more details of the control of trophoblast gene expression to be elucidated.

References

Aboagye-Mathiesen, G., F.D. Toth, M. Zdravkovic, P. Ebbesen (1994) Production of interferons in human placental trophoblast subpopulations and their possible roles in pregnancy. Clin Diagn Lab Immunol *1:* 650–659.

Aplin, J.D. (1991) Implantation, trophoblast differentiation and haemochorial placentation: Mechanistic evidence in vivo and in vitro. J Cell Sci *99:* 681–692.

Aplin, J.D. (1993) Expression of integrin $\alpha_6\beta_4$ in human trophoblast and its loss from extravillous cells. Placenta *14:* 203–215.

Aplin, J.D. (1996) The cell biology of human implantation. Placenta *17:* 269–276.

Aplin, J.D., A.K. Charlton, S. Ayad (1988) An immunohistochemical study of human endometrial extracellular matrix during the menstrual cycle and first trimester of pregnancy. Cell Tissue Res *253:* 231–240.

Aplin, J.D, S.R. Glasser (1994) The interaction of trophoblast with endometrial stroma; in Glasser SR, et al. (eds): Endocrinology of Embryo-Endometrium Interactions. New York, Plenum Press, pp 327–341.

Aplin, J.D., C.J.P. Jones (1989) Extracellular matrix in endometrium and decidua; in Genbacev O, A Klopper, R Beaconsfield (eds): Placenta as a Model and a Source. New York, Plenum Press, pp 115–128.

Aplin, J.D., P. Mylona, C.M. Kielty, S. Ball, H.J. Church, J.D.L. Williams, C.J.P. Jones (1995) Collagen VI and laminin as markers of decidualisation in human endometrial stroma; in Dey SK (ed): Molecular and Cellular Aspects of Periimplantation Processes. Serono Symposium Series. New York, Springer, pp 331–351.

Aplin, J.D., Lj. Vićovac, A. Sattar (1993) Cell interactions in trophoblast invasion; in Soares MJ, S Handwerger, F Talamantes (eds): Trophoblast Cells. Pathways for Maternal-Embryonic Communication, pp 92–108.

Arnholdt, H., J. Diebold, B. Kuhlmann, U. Löhrs (1991) Receptor-mediated processing of epidermal growth factor in the trophoblast of the human placenta. Virchows Arch [B] *61:* 75–80.

Aufderheide, E., P. Ekblom (1988) Tenascin during gut development: Appearance in the mesenchyme, shift in molecular forms and dependence on the epithelial-mesenchymal interactions. J Cell Biol *107:* 2341–2349.

Autio-Harmainen, H., T. Hurskainen, K. Niskasaari, M. Hoyhtya, K. Tryggvason (1992) Simultaneus expression of 70 kDa type IV collagenase and type IV collagen α_1(IV) chain genes by cells of early human placenta and gestational endometrium. Lab Invest *67:* 191–200.

Babawale, M.O., S. van Noorden, M. Pignatelli, G.W.H. Stamp, M.G. Elder, M.H.F. Sullivan (1996) Morphological interaction of human first trimester placental villi co-cultured with decidual explants. Human Reprod *11:* 444–450.

Bass, K.E., D. Morrish, I. Roth, D. Bhardwaj, R. Taylor, Y. Zhou, S.J. Fisher (1994) Human cytotrophoblast invasion is upregulated by epidermal growth factor: Evidence that paracrine factors modify this process. Dev Biol *164:* 550–561.

Beham, A., H. Denk, G. Desoye (1988) The distribution of intermediate filament proteins, actin and desmoplakins in human placental tissue as revealed by polyclonal and monoclonal antibodies. Placenta *9:* 479–492.

Bell, S.C. (1989) Decidualisation and insulin-like growth factor (IGF) binding protein: Implications for its role in stromal cell differentiation and the decidual cell in haemochorial placentation. Human Reprod *4:* 125–130.

Benirschke, K., P. Kaufmann (1990) Pathology of the Human Placenta, ed 2. New York, Springer.

Bergh, P.A., D. Navot (1992) The impact of embryonic development and endometrial maturity on the timing of implantation. Fert Steril *58:* 537–542.

Bergman, D., S.S. Kadner, M.R. Cruz, A.L. Esterman, M.M. Tahery, B.K. Young, T.H. Finlay (1993) Synthesis of α_1-antichymotrypsin and α_1-antitrypsin by human trophoblast. Pediatr Res *34:* 312–317.

Bischof, P., E. Friedli, M. Martelli, A. Campana (1991) Expression of extracellular matrix-degrading metalloproteinases by cultured human cytotrophoblast cells: Effects of cell adhesion and immunopurification. Am J Obstet Gynecol *165:* 1791–1801.

Blankenship, T.N., B.F. King (1994) Identification of 72-kilodalton type IV collagenase at sites of trophoblastic invasion of macaque spiral arteries. Placenta *15:* 177–187.

Bulmer, J.N., P.M. Johnson (1985) Immunohistological characterization of the decidual leukocytic infiltrate related to endometrial gland epithelium in early human pregnancy. Immunology *55:* 35–44.

Bulmer, J.N., C.A. Sunderland (1984) Immunohistological characterization of the lymphoid cell populations in the early human placental bed. Immunology *52:* 349–357.

Bulmer, J.N., S. Thrower, M. Wells (1989) Expression of epidermal growth factor receptor and transferrin receptor by human trophoblast populations. Am J Reprod Immunol *21:* 87–93.

Burdsal, C.A., C.H. Damsky, R.A. Pedersen (1993) The role of E-cadherins in mesoderm differentiation and migration at the mammalian primitive streak. Development *118:* 829–844.

Burgeson, R.E., M. Chiquet, R. Deutzmann, P. Ekblom, J. Engel, H. Kleinman, G.R. Martin, J.-P. Ortonne, M. Paulsson, J. Sanes, R. Timpl, K. Tryggvason, Y. Yamada, P.D. Yurchenco (1994) A new nomenclature for laminins. Matrix Biol *14:* 209–211.

Burrows, T.D., A. King, Y.W. Loke (1993) Expression of integrins by human trophoblast and differential adhesion to laminin or fibronectin. Human Reprod *8:* 475–484.

Burrows, T.D., A. King, Y.W. Loke (1994) Expression of adhesion molecules by endovascular trophoblast and decidual endothelial cells: Implications for vascular invasion during implantation. Placenta *15:* 21–33.

Carmeliet, P., L. Schoonjans, L. Kieckens, B. Ream, J. Degen, R. Bronson, R. De Vos, J.J. van den Oord, D. Collen, R.C. Mulligan (1994) Physiological consequences of loss of plasminogen activator gene function in mice. Nature *368:* 419–424.

Castellucci, M., I. Classen-Linke, J. Muhlhauser, P. Kaufmann, L. Zardi, R. Chiquet-Ehrismann (1991) The human placenta: A model for tenascin expression. Histochemistry *95:* 449–458.

Charnock-Jones, D.S., A.M. Sharkey, C.A. Boocock, A. Ahmed, R. Plevin, N. Ferrara, S.K. Smith (1994a) Vascular endothelial growth factor receptor localisation and activation in human trophoblast and choriocarcinoma cells. Biol Reprod *51:* 524–530.

Charnock-Jones, D.S., A.M. Sharkey, P. Fenwick, S.K. Smith (1994b) Leukaemia inhibitory factor mRNA concentration peaks in human endometrium at the time of implantation and the blastocyst contains mRNA for the receptor at this time. J Reprod Fertil *101:* 421–426.

Chegini, N., Y. Zhao, R.S. Williams, K.C. Flanders (1994) Human uterine tissue throughout the menstrual cycle expresses transforming growth factor-β_1 (TGFβ_1), TGFβ_2, TGFβ_3 and TGFβ type II receptor messenger ribonucleic acid and protein and contains [^{125}I] TGF β_1-binding sites. Endocrinology *135:* 439–449.

Chou, J. (1978) Establishment of clonal human placental cells synthesizing human choriogonadotropin. Proc Natl Acad Sci USA *75:* 1854–1858.

Chumbley, G., A. King, N. Holmes, Y.W. Loke (1993) In situ hybridisation and Northern blot demonstration of HLA-G mRNA in human trophoblast populations by locus-specific oligonucleotide. Hum Immunol *37:* 17–22.

Church, H.J., A.J. Richards, J.D. Aplin (1996a) Laminin in decidua, placenta and choriocarcinoma cells. Trophoblast Res, in press.

Church, H.J., Lj. Vićovac, J.D.L. Williams, N.A. Hey, J.D. Aplin (1996b) Laminins 2 and 4 are expressed by human decidual cells. Lab Invest *74:* 21–32.

Cronier, L., B. Bastide, J.C. Herve, J. Deleze, A. Malassine (1994) Gap junctional communication during human trophoblast differentiation: Influence of hCG. Endocrinology *135:* 402–408.

Cross, J.C., Z. Werb, S.J. Fisher (1994) Implantation and the placenta: Key pieces of the development puzzle. Science *266:* 1508–1518.

Damsky, C.H., M.L. Fitzgerald, S.J. Fisher (1992) Distribution of extracellular matrix components and adhesion receptors are intricately modulated during first trimester cytotrophoblast differentiation along the invasive pathway, in vivo. J Clin Invest *89:* 210–222.

Damsky, C.H., C. Librach, K.-H. Lim, M.L. Fitzgerald, M.T. McMaster, M. Janatpour, Y. Zhou, S.K. Logan, S.J. Fisher (1994) Integrin switching regulates normal trophoblast invasion. Development *120:* 3657–3666.

Damsky, C., A. Sutherland, S. Fisher (1993) Extracellular matrix 5: Adhesive interactions in early mammalian embryogenesis, implantation and placentation. FASEB J *7:* 1320–1329.

Denker, H.-W. (1990) Trophoblast-endometrium interactions at embryo implantation: A cell biological paradox; in Denker H-W, JD Aplin (eds): Trophoblast Invasion and Endometrial Receptivity. New York, Plenum Medical, pp 3–29.

Denker, H.-W. (1993) Implantation: A cell biological paradox. J Exp Zool *266:* 541–558.

Douglas, G.C., B.F. King (1989) Isolation of pure villous cytotrophoblast from term human placenta using immunomagnetic microspheres. J Immunol Methods *119:* 259–268.

Dungy, L.J., T.A. Siddiqi, S. Khan (1991) Transforming growth factor β1 expression during placental development. Am J Obstet Gynecol *165:* 853–857.

Ekström, T.J., L. Holmgren, A. Glaser, R. Ohlsson (1994) IGF2 expression at the embryonic/maternal boundary; in Glasser SR, J Mulholland, A Psychoyos (eds): Endocrinology of Embryo-Endometrial Interactions. New York, Plenum, pp 246–251.

Ellis, S.A., M.S. Palmer, A.J. McMichael (1990) Human trophoblast and the choriocarcinoma cell line BeWo express a truncated HLA class I molecule. J Immunol *144:* 731–735.

Emonard, H., Y. Christiane, M. Smet, J.A. Grimaud, J.M. Foidart (1990) Type IV and interstitial collagenolytic activities in normal and malignant trophoblast cells are specifically regulated by the extracellular matrix. Invas Metast *10:* 170–177.

Enders, A.C. (1968) Fine structure of anchoring villi of the human placenta. Am J Anat *122:* 419–452.

Enders, A.C. (1991) Implantation. Encyclopedia Hum Biol *4.* 423–430.

Enders, A.C., K.C. Lantz, S. Schlafke (1996) Preference of invasive cytotrophoblast for maternal vessels in early implantation in the macaque. Acta Anat *155:* 145–162.

Engvall, E., D. Earwicker, T. Haaparanta, E. Ruoslahti, J.R. Sanes (1990) Distribution and isolation of four laminin variants: Tissue restricted distribution of heterotrimers assembled from five different subunits. Cell Regul *1:* 731–740.

Farmer, D.R., D.M. Nelson (1992) A fibrin matrix modulates the proliferation and hormone secretion and morphologic differentiation of cultured human placental trophoblast. Placenta *13:* 163–177.

Feinberg, R.F., L.-C. Kao, J.E. Haimowitz, J.T. Queenan, T.-C. Wun, J.F. Strauss, H.J. Kliman (1989) Plasminogen activator inhibitor types 1 and 2 in human trophoblasts: PAI-1 is an immunocytochemical marker of trophoblasts. Lab Invest *61:* 20–26.

Feinberg, R.F., H.J. Kliman, C.J. Lockwood (1991) Is oncofetal fibronectin a trophoblast glue for human implantation? Am J Pathol *138:* 537–543.

Feinberg, R.F., H.J. Kliman, C.-L. Wang (1994) Transforming growth factor β stimulates trophoblast oncofetal fibronectin synthesis in vitro: Implications for trophoblast implantation in vivo. J Clin Endocrinol Metab *78:* 1241–1248.

Fernandez, P.I., M.J. Merino, F.F. Nogales, A.S. Charonis, W. Stetler-Stevenson, L. Liotta (1990) Immunohistochemical profile of basement membrane proteins and 72kDa type IV collagenase in the implantation placental site. Lab Invest *66:* 572–579.

Filla, M.S., C.X. Zhang, K.L. Kaul (1993) A potential transforming growth factor alpha/epidermal growth factor receptor autocrine circuit in placental cytotrophoblasts. Cell Growth Differen *4:* 387–393.

Fisher, S.J., T.-Y. Cui, L. Zhang, L. Hartman, K. Grahl, Z. Guo-Yang, J. Tarpey, C.H. Damsky (1989) Adhesive and degradative properties of human placental cytotrophoblast cells in vitro. J Cell Biol *109:* 891–902.

Fisher, S.J., M.S. Leitch, M.S. Kantor, C.B. Basbaum, R.H. Kramer (1985) Degradation of the extracellular matrix by the trophoblastic cells of first trimester human placenta. J Cell Biochem *27:* 31–41.

Frank, H.-G., F. Malekzadeh, S. Kertchanska, C. Crescimanno, M. Castelluci, I. Lang, G. Desoye, P. Kaufmann (1994) Immunohistochemistry of two different types of placental fibrinoid. Acta Anat *150:* 55–68.

Garcia-Lloret, M.I., D.W. Morrish, T.G. Wegmann, L. Honore, A.R. Turner, L.J. Guilbert (1994) Demonstration of functional cytokine-placental interactions: CSF-1 and GM-CSF stimulate human cytotrophoblast differentiation and peptide hormone secretion. Exp Cell Res *214:* 46–54.

Genbačev, O., S.A. Schubach, R.K. Miller (1992) Villous culture of first trimester human placenta: Model to study extravillous trophoblast differentiation. Placenta *13:* 439–461.

Glaser, A., H. Luthman, I. Stern, R. Ohlsson (1992) Spatial distribution of active genes implicated in the regulation of IGF stimulatory loops in human decidual and placental tissue of first trimester pregnancy. Mol Reprod Dev *33:* 7–15.

Gosseye, S., H. Fox (1984) An immunological comparison of the secretory capacity of villous and extravillous trophoblast in the human placenta. Placenta *5:* 329–348.

Graham, C.H., T.S. Hawley, R.G. Hawley, J.R. MacDougall, R.S. Kerbel, N. Khoo, P.K. Lala (1993) Establishment and characterization of first trimester human trophoblast cells with extended lifespan. Exp Cell Res *206:* 204–211.

Graham, C.H., P.K. Lala (1991) Mechanism of control of trophoblast invasion in situ. J Cell Physiol *148:* 228–234.

Graham, C.H., J.J. Lysiak, K.R. McCrae, P.K. Lala (1992) Localisation of TGFβ at the human fetal-maternal interface: Role in trophoblast growth and differentiation. Biol Reprod *46:* 561–572.

Haining, R.E.B., J.P. Schofield, D.S.C. Jones, J. Rajput-Williams, S.K. Smith (1991) Identification of mRNA for EGF and TGFα present in low copy number of human endometrium and decidua using RT-PCR. J Mol Endocrinol *6:* 207–214.

Hamlin, G.P., M.J. Soares (1995) Regulation of deoxyribonucleic acid synthesis in proliferating and differentiating trophoblast cells: Involvement of transferrin, transforming growth factor β, and tyrosine kinases. Endocrinology *136:* 322–331.

Hay, E.D. (1995) An overview of epithelio-mesenchymal transformation. Acta Anat *154:* 8–20.

Haynes, M.K., L.G. Jackson, R.S. Tuan, K.J. Shepley, J.B. Smith (1993) Cytokine production in first trimester chorionic villi: Detection of mRNAs and protein products in situ. Cell Immunol *151:* 300–308.

Hertig, A.T., J. Rock, J. Adams (1956) A description of 34 human ova within the first 17 days of development. Am J Anat *98:* 325–494.

Hofmann, G.E., G.M. Horowitz, R.T. Scott, D. Navot (1993) TGFα in human implantation trophoblast: Immunohistochemical evidence for autocrine/paracrine function. J Clin Endocrinol Metab *76:* 781–785.

Hohn, H.-P., L.R. Boots, H.-W. Denker, M. Hook (1993) Differentiation of human trophoblast cells in vitro stimulated by extracellular matrix. Trophoblast Res *7:* 181–200.

Hohn, H.-P., H.-W. Denker (1994) The role of cell shape for differentiation of choriocarcinoma cells on extracellular matrix. Exp Cell Res *215:* 40–50.

Holmgren, L., L. Claesson-Welsh, C.H. Heldin, R. Ohlsson (1992) The expression of PDGF a and b receptors in subpopulations of PDGF-producing cells implicates autocrine stimulatory loops in the control of proliferation in cytotrophoblast that have invaded the maternal endometrium. Growth Factors *6:* 219–231.

Holmgren, L., A. Glaser, S. Pfeiffer-Olsson, R. Ohlsson (1991) Angiogenesis during human extra-embryonic development involves the spatio-temporal control of PDGF ligand and receptor gene expression. Development *113:* 749–754.

Hoshina, M., M. Boothby, I. Boime (1982) Cytological localization of chorionic gonadotropin and placental lactogen mRNAs during development of the human placenta. J Cell Biol *93:* 190–198.

Hoshina, M., M. Boothby, R. Hussa, R. Pattillo, H.M. Camel, I. Boime (1983) Linkage of human chorionic gonadotrophin and placental lactogen biosynthesis to trophoblast differentiation and tumorigenesis. Placenta *6:* 163–172.

Hu, X.L., Y. Yang, J.S. Hunt (1992) Differential distribution of IL-1α and IL-1β proteins in human placentas. J Reprod Immunol *22:* 257–268.

Irving, J.A., P.K. Lala (1995) Functional role of cell surface integrins on human trophoblast cell migration: regulation by TGFβ, IGF-II and IGFBP-1. Exp Cell Res *217:* 419–427.

Irving, J.A., J.J. Lysiak, S. Hearn, P.K. Lala (1995) Characterisation of trophoblast cells migrating from first trimester chorionic villous explants and propagated in culture. Placenta *16:* 413–433.

Jaffe, R.B., P.A. Lee, A.R. Midgley, Jr. (1969) Serum gonadotropins before, at the inception of and following human pregnancy. J Clin Endocrinol Metab *29:* 1281–1283.

Jokhi, P.P., G. Chumbley, A. King, L. Gardner, Y.W. Loke (1993) Expression of the colony stimulating factor-1 receptor (c-fms product) by cells at the human uteroplacental interface. Lab Invest *68:* 308–320.

Jokhi, P.P., A. King, P.T. Jubinsky, Y.W. Loke (1994a) Demonstration of the low affinity alpha subunit of the granulocyte-macrophage colony-stimulating factor receptor (GM-CSF-Rα) on human trophoblast and uterine cells. J Reprod Immunol *26:* 147–164.

Jokhi, P.P., A. King, Y.W. Loke (1994b) Reciprocal expression of epidermal growth factor receptor (EGF-R) and c-erbB2 by non-invasive and invasive human trophoblast populations. Cytokine *6:* 433–442.

Jokhi, P.P., A. King, Y.W. Loke (1994c) Production of GM-CSF by human trophoblast cells and by decidual large granulated lymphocytes. Hum Reprod *9:* 1660–1669.

Kauma, S., D. Matt, S. Strom, D. Eierman, T. Turner (1990) IL-1β, HLA-DRα and TGFβ expression in endometrium, placenta and placental membranes. Am J Obstet Gynecol *163:* 1430–1437.

King, B.F., T.N. Blankenship (1994) Ultrastructure and development of a thick basment membrane-like layer in the anchoring villi of macaque placentas. Anat Rec *238:* 498–506.

Kirszenbaum, M., P. Moreau, E. Gluckman, J. Dausset, E. Carosella (1994) An alternatively spliced form of HLA-G in human trophoblasts and evidence for the presence of HLA-G transcript in adult lymphocytes. Proc Natl Acad Sci USA *91:* 4209–4213.

Kliman, H.J., R.F. Feinberg (1992) Differentiation of the trophoblast; in Barnea ER, J Hustin, E Jauniaux (eds): The First Twelve Weeks of Gestation. Berlin, Springer, pp 3–25.

Kliman, H.J., J.E. Nestler, E. Sermasi, J.M. Sanger, J.F. Strauss III (1986) Purification, characterization and in vitro differentiation of cytotrophoblasts from term placentae. Endocrinology *118:* 1567–1582.

Kojima, K., H. Kanzaki, M. Iwai, H. Hatayama, M. Fujimoto, T. Inoue, K. Horie, H. Nakayama, J. Fujita, T. Mori (1994) Expression of leukemia inhibitory factor in human endometrium and placenta. Biol Reprod *50:* 882–887.

Korhonen, M., J. Ylanne, L. Laitinen, H.M. Cooper, V. Quaranta, I. Virtanen (1991) Distribution of the $\alpha_1-\alpha_6$ integrin subunits in human developing and term placenta. Lab Invest *65:* 347–356.

Kurman, R.J., C.S. Main, H.-C. Chem (1984) Intermediate trophoblast: a distinctive form of trophoblast with specific morphological, biochemical and functional features. Placenta *5:* 349–370.

Lala, P.K., C.H. Graham (1990) Mechanisms of trophoblast invasiveness and their control: The role of proteases and protease inhibitors. Cancer Metast Rev *9:* 369–379.

Larue, L., M. Ohsugi, J. Hirchenhain, R. Kemler (1994) E-cadherin null mutant embryos fail to form a trophectoderm epithelium. Proc Natl Acad Sci USA *91:* 8263–8267.

Lehtonen, E., V.P. Lehto, T. Vartio, R.A. Badley, I. Virtanen (1983) Expression of cytokeratin polypeptides in mouse oocytes and preimplantation embryos. Dev Biol *100:* 158–165.

Leivo, I., P. Laurila, T. Wahlström, E. Engvall (1989) Expression of merosin, a tissue-specific basement membrane protein, in the intermediate trophoblast cells of choriocarcinoma and placenta. Lab Invest *60:* 783–790.

Letterio, J.J., A.G. Geiser, A.B. Kulkarni, N.S. Roche, M.B. Sporn, A.B. Roberts (1994) Maternal rescue of TGFβ_1 null mice. Science *264:* 1936–1938.

Librach, C.L., S.L. Feigenbaum, K.E. Bass, T.Y. Cui, N. Verastas, Y. Sadovsky, J.P. Quigley, D.L. French, S.J. Fisher (1994) Interleukin 1β regulates human cytotrophoblast metalloproteinase activity and invasion in vitro. J Biol Chem *269:* 17125–17131.

Librach, C.L., Z. Werb, M.L. Fitzgerald, K. Chiu, N.M. Corwin, R.A. Esteves, D. Grobelny, R. Galardy, C.H. Damsky, S.J. Fisher (1991) 92kDa type IV collagenase mediates invasion of human cytotrophoblasts. J Cell Biol *113:* 437–449.

Logan, S.K., S.J. Fisher, C.H. Damsky (1992) Human placental cells with temperature-sensitive simian virus 40 are immortalised and mimic the phenotype of invasive cytotrophoblasts at both permissive and nonpermissive temperatures. Cancer Res *52:* 6001–6009.

Loke, Y.W., B.H. Butterworth (1987) Heterogeneity of human trophoblast populations; in Gill TJ, TG Wegman (eds): Immunoregulation and Fetal Survival. London, Oxford University Press, pp 197–209.

Loke, Y.W., A. King, L. Gardner, N.P. Carter (1992) Evidence for the expression of GM-CSF receptors by human first trimester extravillous trophoblast and its response to this cytokine. J Reprod Immunol *22:* 33–45.

Lysiak, J.J., V.K. Han, P.K. Lala (1993) Localization of transforming growth factor α in the human placenta and decidua: Role in trophoblast growth. Biol Reprod *49:* 885–894.

Mackay, A.R., D.E. Gomez, D.W. Cottam, R.C. Rees, A.M. Nason, U.P. Thorgeirsson (1993) Identification of the 72-kDa (MMP-2) and 92-kDa (MMP-9) gelatinase/type IV collagenase in preparations of laminin and Matrigel. Biotechniques *15:* 1048–1051.

McMaster, M.T., C.L. Librach, Y. Zhou, K.H. Lim, M.J. Janatpour, R. DeMars, S. Kovats, C. Damsky, S.J. Fisher (1995) Human placental HLA-G expression is restricted to differentiated cytotrophoblasts. J Immunol *154:* 3771–3778.

Maquoi, E., M. Polette, B. Nawrocki, P. Bischof, A. Noel, A. Pintiaux, M. Santavicca, J.-P. Schapps, R. Pijnenborg, P. Birembaut, J.-M. Foidart (1996) Expression of stromelysin 3 in the human placenta and placental bed. Placenta, in press.

Maruo, T., H. Matsuo, K. Murata, M. Mochzuki (1992) Gestational age-dependent dual action of epidermal growth factor on human placenta early in gestation. J Clin Endocrinol Metab *75:* 1362–1367.

Masuhiro, K., N. Matsuzaki, E. Nishino, T. Taniguchi, T. Kameda, Y. Li, F. Saji, O. Tanizawa (1992) Trophoblast-derived IL-1 stimulates the release of hCG by activating IL-6 and IL-6 receptor system in first trimester human trophoblasts. J Clin Endocrinol Metab *72:* 594–601.

Matsuzaki, N., Y. Li, K. Masuhiro, T. Jo, K. Shimoya, T. Taniguchi, F. Saji, O. Tanizawa (1992) Trophoblast-derived TGF-β_1 suppresses cytokine-induced but not gonadotropin-releasing hormone-induced release of hCG by normal human trophoblasts. J Clin Endocrinol Metab *74:* 211–216.

Mitchell, M.D., M.S. Trautman, D.J. Dudley (1993) Cytokine networking in the placenta. Placenta *14:* 249–275.

Mitchell, E.J., L. Fitzgibbon, M.D. O'Connor-McCourt (1994) Subtypes of betaglycan and and of type I and type II TGF-b receptors with different affinities for TGFβ_1 and TGFβ_2 are exhibited by human placental trophoblast cell. J Cell Physiol *150:* 334–343.

Mitchell, E.J., M.D. O'Connor-McCourt (1991) A TGFβ receptor from human placenta exhibits a greater affinity for TGFβ_2 than for TGFβ_1. Biochemistry *30:* 4350–4356.

Moll, U.M., B.L. Lane (1990) Proteolytic activity first trimester human placenta: Localization of interstitial collagenase in villous and extravillous trophoblast. Histochemistry *94:* 555–560.

Morrish, D.W., D. Bhardwaj, L.K. Dabbagh, H. Marusyk, O. Siu (1987) Epidermal growth factor induces differentiation and secretion of human chorionic gonadotropin and placental lactogen in normal human placenta. J Clin Endocrinol Metab *65:* 1282–1290.

Morrish, D.W., D. Bhardwaj, M.T. Paras (1991) TGFb$_1$ inhibits placental differentiation and human chorionic gonadotropin and placental lactogen secretion. Endocrinology *120:* 22–26.

Mühlhauser, J., C. Crescimanno, M. Kasper, D. Zaccheo, M. Castellucci (1995) Differentiation of human trophoblast populations involves alterations in cytokeratin patterns. J Histochem Cytochem *43:* 579–589.

Mühlhauser, J., C. Crescimanno, P. Kaufmann, H. Hofler, D. Zaccheo, M. Castellucci (1993) Differentiation and proliferation patterns in human trophoblast revealed by c-erbB-2 oncogene product and EGF-R. J Histochem Cytochem *41:* 165–173.

Nelson, D.M., E.C. Crouch, E.M. Curran, D.R. Farmer (1990) Trophoblast interaction with fibrin matrix: Epithelization of perivillous fibrin deposits as a mechanism for villous repair in the human placenta. Am J Pathol *136:* 855–865.

Ohlsson, R., L. Holmgren, A. Glaser, A. Szpecht, S. Pfeifer-Ohlsson (1989) Insulin-like growth factor 2 and short range stimulatory loops in control of human placental growth. EMBO J *8:* 1993–1999.

Ohno, M., A. Martinez-Hernandez Ohno, N.A. Kefalides (1986) Laminin M is found in placental basement membranes, but not in basement membranes of neoplastic origin. Connect Tiss Res *15:* 199–207.

Pijnenborg, R. (1994) Trophoblast invasion. Reprod Med Rev *3:* 53–73.

Pijnenborg, R., G. Dixon, W.B. Robertson, I. Brosens (1980) Trophoblastic invasion of human decidua from 8 to 18 weeks of pregnancy. Placenta *1:* 3–19.

Polette, M., B. Nawrocki, A. Pintiaux, C. Massenat, E. Maquoi, L. Volders, J.P. Schnapps, P. Birembaut, J.M. Foidart (1994) Expression of gelatinases A and B and their tissue inhibitors by cells of early and term human placenta and gestational endometrium. Lab Invest *71:* 838–846.

Psychoyos, A. (1988) The 'implantation window': Can it be enlarged or displaced? In: Human Reproduction. Excerpta Med Int Congr Ser *768:* 231–232.

Queenan, J.T., L.-C. Kao, C.E. Arboleda, A. Ulloa-Aguirre, T.G. Golos, B. Cines, J.F. Strauss III (1987) Regulation of urokinase-type plasminogen activator production by cultured human cytotrophoblasts. J Biol Chem *262:* 10903–10906.

Redman, C.W.G., A.J. McMicael, G.M. Stirrat, C.A. Sunderland, A. Ting (1984) Class I major histocompatibility complex antigens on human extravillous trophoblast. Immunology *52:* 457–468.

Riddick, D.H., A.A. Luciano, W.F. Kusmik, I.A. Maslar (1978) De novo synthesis of prolactin by human decidua. Life Sci *23:* 1913–1922.

St. Jacques, S., M. Forte, S.J. Lye, M. Letarte (1994) Localisation of endoglin, a TGFβ binding protein and of CD44 and integrins in placenta during the first trimester of pregnancy. Biol Reprod *51:* 405–413.

Sakbun, V., S.M. Ali, Y.A. Lee, C.S. Jara, G.D. Bryant-Greenwood (1990) Immunocytochemical localization and messenger ribonucleic acid concentrations for human placental lactogen in amnion, chorion, decidua and placenta. Am J Obstet Gynecol *162:* 1310–1317.

Sasagawa, M., T. Yamazaki, M. Endo, K. Kanazava, S. Takeuchi (1987) Immunohistochemical localization of HLA antigens and placental proteins (αhCG, βhCG, CTP, hPL and SP1) in villous and extravillous trophoblast in normal human pregnancy: A distinctive pathway of differentiation of extravillous trophoblast. Placenta *8:* 515–528.

Schaaps, J.P., J. Hustin (1988) In vivo aspect of the materno-trophoblastic border during the first trimester gestation. Trophoblast Res *3:* 3–48.

Schatz, F., C. Papp, E. Toth-Pal, C.J. Lockwood (1994) Ovarian steroid-modulated stromelysin-1 expression in human endometrial stromal and decidual cells. J Clin Endocrinol Metab *78:* 1467–1472.

Schlafke, S., A.C. Enders (1975) Cellular basis of interaction between trophoblast and uterus at implantation. Biol Reprod *12:* 41–65.

Schmidt, C., F. Bladt, S. Goedecke, V. Brinkmann, W. Zschiesche, M. Sharpe, E. Gherardi, C. Birchmeier (1995) Scatter factor/HGF is essential for liver development. Nature *373:* 699–702.

Selick, C.E., G.M. Horowitz, M. Gratch, R.T. Scott, Jr., D. Navot, G.E. Hofmann (1994) Immunohistochemical localization of transforming growth factor β in human implantation sites. J Clin Endocrinol Metab *78:* 592–596.

Shorter, S.C., P.M. Starkey, B.L. Ferry, L.M. Clover, I.L. Sargent, C.W.G. Redman (1993) Antigenic heterogeneity of human cytotrophoblast and evidence for the transient expression of MHC class I antigens distinct from HLA-G. Placenta *14:* 571–582.

Simon, C., A. Frances, G.N. Piquette, I. El Danasouri, G. Zurawski, W. Dang, M.L. Polan (1994) Embryonic implantation in mice is blocked by IL-1 receptor antagonist. Endocrinology *134:* 521–528.

Starkey, P.M. (1993) The decidua and factors controlling placentation; in Redman CWG, IL Sargent, PM Starkey (eds): The Human Placenta. Oxford, Blackwell Scientific Publication, pp 362–413.

Starkey, P.M., I.L. Sargent, C.W.G. Redman (1988) Cell populations in human early pregnancy decidua: characterization and isolation of large granular lymphocytes by flow cytometry. Immunology *65:* 129–134.

Stewart, C.L., P. Kaspar, L.J. Brunet, H. Bhatt, I. Gadi, F. Kontgen, S.J. Abbondanzo (1992) Blastocyst implantation depends on maternal expression of leukaemia inhibitory factor. Nature *359:* 76–79.

Strickland, S., W.G. Richards (1992) Invasion of the trophoblasts. Cell *71:* 355–357.

Tabibzadeh, S. (1994) Role of cytokines in endometrium and at the fetomaternal interface. Reprod Med Rev *3:* 11–28.

Teasdale, F., E.W. Adcock III, C.S. August, S. Cox, F.C. Battaglia, M.A. Naughton (1973) Human chorionic gonadotropin: Inhibitory effect on mixed lymphocyte culture. Gynecol Invest *4:* 263–269.

Uehara, Y., O. Minowa, C. Mori, K. Shiota, J. Kuno, T. Noda, N. Kitamura (1995) Placental defect and embryonic lethality in mice lacking scatter factor. Nature *373:* 702–705.

Vićovac, Lj., C.J.P. Jones, J.D. Aplin (1995a) Trophoblast differentiation during formation of anchoring villi in a model of the early human placenta in vitro. Placenta *16:* 41–56.

Vićovac, Lj., C.J.P. Jones, H.J. Church, J.D. Aplin (1995b) Factors controlling trophoblast entry into the extravillous lineage (abstract). Placenta *16:* A74.

Vićovac, Lj., N. Papic, J.D. Aplin (1993) Tissue interactions in first trimester trophoblast-decidua cocultures. Trophoblast Res *7:* 223–236.

Vince, G.S., S.C. Shorter, P.M. Starkey (1992) Localization of tumour necrosis factor production in cells at the materno/fetal interface in human pregnancy. Clin Exp Immunol *88:* 174–180.

Vukicevic, S., H.K. Kleinman, F.P. Luyten, A.B. Roberts, N.S. Roche, A.H. Reddi (1992) Identification of multiple active growth factors in basement membrane Matrigel suggests caution in interpretation of cellular activity related to extracellular matrix components. Exp Cell Res *202:* 1–8.

Wewer, U.M., M. Faber, L.A. Liotta, R. Albrechtsen (1985) Immunochemical and ultrastructural assessment of the nature of the pericellular basement membrane of human decidual cells. Lab Invest *53:* 624–633.

Wolf, H.K., R. Zarnegar, L. Oliver, G.K. Michalopoulos (1991) HGF in human placenta and trophoblastic disease. Am J Pathol *138:* 1035–1043.

Yagel, S., R.F. Casper, W. Powell, R.S. Parhar, P.K. Lala (1989) Characterization of pure human first-trimester cytotrophoblast cells in long-term culture: Growth pattern, markers, and hormone production. Am J Obstet Gynecol *160:* 938–945.

Yagel, S., T.G. Geva, H. Solomon, S. Shimonovitz, R. Reich, Z. Finci-Yerheskel, M. Mayer, A. Milwidsky (1993) High levels of human chorionic gonadotrophin retard first trimester trophoblast invasion in vitro by decreasing urokinase plasminogen activator and collagenase activities. J Clin Endocrinol Metab *77:* 1506–1511.

Yagel, S., R.S. Parhar, J.J. Jeffrey, P.K. Lala (1988) Normal nonmetastatic human trophoblast cells share in vitro invasive properties of malignant cells. J Cell Physiol *136:* 455–462.

Yang, Y., K.K. Yelavarthi, H.L. Chen, J.l. Pace, P.F. Terranova, J.S. Hunt (1993) Molecular, biochemical, and functional characteristics of tumor necrosis factor α produced by human placental cytotrophoblastic cells. J Immunol *150:* 5614–5624.

Yen, S.S.C., R.B. Jaffe (1991) Reproductive Endocrinology, ed 3. Philadelphia, Saunders, chap 26.

Yui, J., M. Garcia-Lloret, A.J. Brown, R.C. Berdan, D.W. Morrish, T.G. Wegmann, L.J. Guilbert (1994) Functional, long-term cultures of human term trophoblasts purified by column-elimination of CD-9 expressing cells. Placenta *15:* 231–246.

Zhang, Z., C. Funk, D. Roy, S. Glasser, J. Mulholland (1994) Heparin-binding EGF-like growth factor is differentially regulated by progesterone and estradiol in rat uterine epithelial and stromal cells. Endocrinology *124:* 1089–1094.

Zhou, Y., C.H. Damsky, K. Chiu, J.M. Roberts, S.J. Fisher (1993) Preeclampsia is associated with abnormal expression of adhesion molecules by invasive cytotrophoblast. J Clin Invest *91:* 950–960.

Acta Anatomica

Acta Anat 1996;156:217–226

C. Birchmeier[a]
W. Birchmeier[b]
B. Brand-Saberi[c]

Departments of
[a] Medical Genetics, and
[b] Cell Biology, Max-Delbrueck Zentrum für molekulare Medizin, Berlin,
[c] Institut für Anatomie II der Universität Freiburg, Germany

Epithelial-Mesenchymal Transitions in Cancer Progression

Key Words
C-met
SF/HGF
E-cadherin
Development
Morphogenesis

Abstract

Epithelial cells are the most important cell type in the development of human malignancies. More than 90% of all malignant tumors are carcinomas, and thus of epithelial origin. Aberrant growth and the ability to invade the underlying tissues are intrinsic properties of the fatally altered cells. Multiple genetic alterations that can influence growth and genetic stability of the carcinoma cells have been characterised during tumor progression. Loss of epithelial morphology and the acquisition of mesenchymal characteristics are typical for carcinoma cells late in tumor progression and correlate with metastatic potential. In vitro, epithelial-mesenchymal transitions can be induced by interference with the integrity of the adherens junction, by signalling via tyrosine kinases, and by oncogene expression. In carcinoma cells, loss or downregulation of *E-cadherin* expression are frequently observed in carcinomas, and correlate with the malignancy of the tumor. In general, this change in expression is regulated at the transcriptional level. However, tumor types or cell lines exist which show mesenchymal characteristics but nevertheless express E-cadherin protein or mRNA. A more-detailed analysis demonstrated that other mechanisms that interfere with E-cadherin-mediated cell adhesion can be operative. Mutations in the *E-cadherin* gene and loss or mutation of the second, intact copy as well as mutation in the *catenin* genes, which encode proteins that interact with the cytoplasmic portion of E-cadherin, can be observed. In addition, transient or unregulated phosphorylation by receptor tyrosine kinases or oncogenic tyrosine kinases, respectively, can interfere with the epithelial morphology and induce a mesenchymal conversion. Since tyrosine phosphorylation of β-catenin correlates with the epithelial-mesenchymal transition that is observed, E-cadherin-mediated cell adhesion might be modulated by such a mechanism. Interestingly, the same molecules implicated in the control of malignant properties turn out to play fundamental roles in the control of normal epithelial growth, differentiation and morphogenesis.

Structural and Molecular Characteristics of Epithelia

Typically, epithelia form continuous sheets of tightly adhering cells that are normally cuboidal in shape. In cell culture, epithelia grow in aggregates and show little cell motility. Characteristic are specialised organelles, tight junctions, adherens junctions and desmosomes that are responsible for the tight intercellular contacts between the individual cells. A consequence of the laterally located junctional complexes is the cell polarity characteristic for epithelial cells. Distinct proteins are expressed on the basolateral or apical

KARGER
E-Mail karger@karger.ch
Fax + 41 61 306 12 34
http://www.karger.ch

0001–5180/96/1563–0217$12.00/0

Carmen Birchmeier
Department for Medical Genetics
Max-Delbrueck Zentrum für molekulare Medizin
Robert-Roessle-Strasse 10
D–13122 Berlin (Germany)

surface and free diffusion of membrane proteins to all surfaces of the cells is inhibited [Rodriguez-Boulan and Nelson, 1989; Schwarz et al., 1990; Birchmeier and Behrens, 1994]. Polar epithelial cells have evolved special mechanisms that allow the transport of membrane proteins to either the apical or basolateral surface [Simons and Wandinger-Ness, 1990].

Adherens junctions [Geiger et al., 1995] are specialised structures containing the transmembrane cell adhesion molecule E-cadherin that recognises and binds E-cadherin present on the neighbouring cells in a Ca^{2+}-dependent manner. E-cadherin is a 120-kD transmembrane glycoprotein, of which an extracellular 80-kD tryptic fragment can be released in the presence of Ca^{2+}. The cDNAs of mouse, chicken and human *E-cadherin* have been characterised; they code for a signal peptide at the amino-terminus, a large extracellular sequence with four repeated domains important in Ca^{2+} binding, a single transmembrane sequence, and a 15-kD cytoplasmic domain [Takeichi, 1991]. E-cadherin is the prototype of a family of Ca^{2+}-dependent cell adhesion molecules; close relatives are N-cadherin (neurones) and P-cadherin (placenta and epithelia). Further members are M-cadherin (muscle), OB-cadherin (osteoblasts), LI-cadherin (liver, intestine), as well as the desmosomal proteins desmoglein and desmocollin, and others [Takeichi, 1991; Donalis et al., 1991; Buxton and Magee, 1992; Kemler, 1992; Sano et al., 1993; Okazaki et al., 1994; Berndorff et al., 1994].

The cytoplasmic portion of E-cadherin interacts with the α-, β- and γ-catenins (plakoglobin) [Takeichi et al., 1991; Kemler, 1993; Tsukita et al., 1993; Troyanovsky et al., 1994]. In addition, other cytoskeletal proteins (vinculin, α-actinin, radixin, ezrin and moesin) are located on the cytoplasmatic side of the junctional complex that anchor the actin-containing filaments [Tsukita et al., 1992]. The integrity of this junctional complex is critical for the maintenance of the functional characteristics of epithelia. Its disruption leads to the dissociation of epithelial sheets, a change in cell morphology to a more flattened shape, and to an increased motility of the cells (fig. 1). The cells thus acquire characteristics of mesenchymal cells (see also below). A controlled loss of epithelial character concomitant with dissociation and increased motility of the cells is a prerequisite for normal morphogenic processes and occurs, for example, during mesoderm formation [Viebahn, 1995] or during the formation of migratory myogenic precursor cells. A similar but uncontrolled epithelial mesenchymal conversion is observed during progression of human tumors [Valles et al., 1991; Birchmeier and Behrens, 1994]. There, loss of epithelial differentiation and the acquisition of mesenchymal characteristics, like the ability to move, correlates with the malignancy of carcinoma cells (fig. 1) [Birchmeier and Behrens, 1994].

Hemidesmosomes are located on the basal surface of epithelia, and are junctions contacting the basement membrane [Garrod, 1993]. The basement membrane separates the epithelial cell compartment from underlying mesenchymal cells and is formed by both cell types. Characteristic constituents of basement membranes are laminin, collagen IV, nidogen/entactin and basement proteoglycan. The mesenchymal reticular lamina which lies below the basement membrane contains collagen type I and III as well as fibronectin [Timpl et al., 1989]. The epithelial receptors for basement membrane components are located on the basal surface of the plasma membrane. Expressed on the epithelial cells are, for example, the integrins α_6/β_1 and α_6/β_4 which are receptors for laminin, and the integrin α_1/β_1 which is a receptor for collagen [Sonnenberg, A., et al., 1991]. The hemidesmosomes are enriched for the integrin receptor α_6/β_4. A variety of experimental evidence is accumulating that demonstrates a function of integrins not only in cell adhesion, but also in the transduction of signals that can influence growth and motility of cells [Juliano and Varner, 1993; Garrod and Humphries, 1995; Geiger et al., 1995].

In addition, various receptors with tyrosine kinase activity are expressed exclusively or predominantly on epithelial cells, like c-Ros, c-Met, c-Neu, c-Ret or the receptor for the keratinocyte growth factor (KGF) [Kokai et al., 1987; Quirke et al., 1989; Mori et al., 1989; Press et al., 1990; Sonnenberg, E., et al., 1991, 1993; Orr-Urtreger et al., 1993; Pachnis et al., 1993]. Since most of these receptors were identified or isolated via their transforming activity in NIH3T3 cells, they can clearly give mitogenic signals [Aaronson, 1991]. However, experiments with epithelial cells in culture have shown that the receptors do not only control growth, but they can also influence motility and shape of epithelia as well as differentiation or morphogenesis [Birchmeier et al., 1993]. Genetic analysis in the mouse also demonstrates the pivotal role of these paracrine signalling molecules in normal development and physiology of epithelia [Werner et al., 1993; Schuchardt et al., 1994; Werner et al., 1994; Peters et al., 1994; Schmidt et al., 1995]. Moreover, mutations giving deregulated expression of such receptors as well as the formation of autocrine loops are observed in carcinoma cells and contribute to the formation of tumors [Aaronson, 1991].

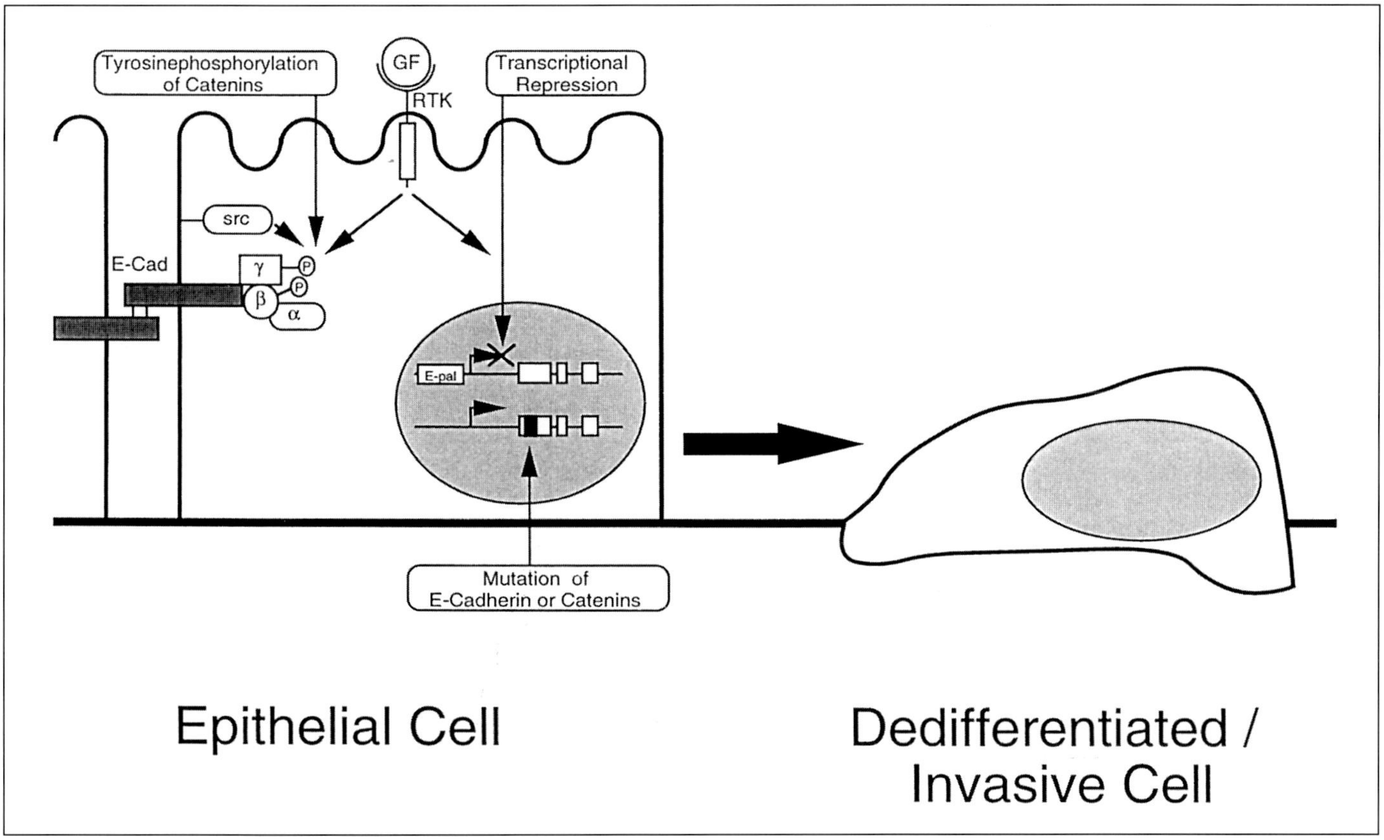

Fig. 1. Mechanisms leading to junctional deficiency. Various molecular mechanisms can interfere with the integrity of the adherens junction and can induce epithelial cells (left) to assume mesenchymal characteristics (right), which correlates with a dedifferentiation of the cells and increased metastatic potential. Loss of E-cadherin, either by downregulation of expression or by mutations of the structural gene are observed, and tyrosine phosphorylation of β-catenin by src or receptor tyrosine kinases (RTK) that recognise specific growth factors (GF) can also interfere with E-cadherin-mediated cell adhesion.

Structural and Molecular Characteristics of Mesenchyme

In contrast to epithelial cells, which display strong intercellular adhesion, basolateral polarization and which are largely nonmobile, mesenchymal cells are generally loosely associated, and surrounded by extracellular matrix. The major adhesive interactions of mesenchymal cells occur with the cell substrate [Hay, 1990]. Integrin molecules on the cell surface mediate the adhesion to extracellular matrix proteins of the connective tissues, for instance to fibronectin, tenascin, or collagen [Hynes, 1992, for a review]. Mesenchymal cells produce particular integrin receptors, for instance $\alpha_5\beta_1$, which enables them to adhere to the extracellular matrix within the connective tissue [Hynes, 1992]. Mesenchymal cells also express specific ligands for various epithelial receptor tyrosine kinases, like SF/HGF or KGF [Werner et al., 1992; Sonnenberg et al., 1993; Mason et al., 1994], and thereby play an important role in the control of epithelial growth, morphogenesis and differentiation [cf. Birchmeier and Birchmeier, 1993].

Molecular Control of Epithelial-Mesenchymal Transitions

Epithelial-mesenchymal transitions (EMTs) can be observed during development, for example in mesoderm formation, when the first mesenchymal cells are formed. Although this demonstrates the intrinsic ability of epithelia to change their cellular characteristics, epithelial-mesenchymal conversions in development occur in a temporally and spatially restricted and controlled manner [Hay, 1995]. The signals responsible for mesoderm induction have been extensively studied, and fibroblast growth factors (FGFs), activins and noggin have been implicated [Beddington and Smith, 1993; Harland et al., 1994]. In contrast, epithelial-mesenchymal conversions can also be induced in cell culture [Stoker et al., 1987]; the epithelial cells that respond appear well differentiated and can derive from tissues where EMTs are not observed under normal, i.e. non-pathological, conditions.

Transitions of epithelia to mesenchyme can be induced in vitro by interfering with the functional integrity of the adherens junctions (fig. 1). The first experiments that demonstrated this were performed by the addition of antibodies that interfere with the function of E-cadherin [Behrens et al., 1989]. The epithelial cells treated in this manner change their shape, are dissociated and resemble fibroblasts in morphology. In addition, they acquire an increased motility and become invasive. When cocultured with heart tissue, the polar Madine Darby kidney (MDCK) cells form a single layer of epithelial sheets that surrounds the tissue. In the presence of anti-E-cadherin antibodies, the cells become motile and invade the heart tissue. The presence of functional adherens junctions is therefore necessary for the epithelial morphology; its disruption leads to an EMT paralleled by an increase in motility and invasiveness of the cells [Behrens et al., 1989; Frixen et al., 1991; Vleminckx et al., 1991].

More recently, the importance of E-cadherin in the maintenance of epithelial cell morphology was also demonstrated by the use of transdominant E-cadherin mutant proteins expressed in epithelial cells [Kintner, 1992; Fujimori and Takeichi, 1993]. Truncated variants of E-cadherin lacking large parts of the extracellular domain but with an intact transmembrane and cytoplasmic sequence are unable to bind to E-cadherin expressed on neighbouring cells. However, they can interact with the cytoplasmatically located catenins. When expressed in large amounts, they can deplete the cellular pool of catenins and thus interfere with normal E-cadherin-catenin interactions. Expressed on cells with epithelial morphology, transdominant E-cadherin induces a fibroblast morphology and an increase in cellular motility.

E-cadherin is not only important for the maintenance of morphology and adhesion in differentiated epithelia, but also for the acquisition of epithelial characteristics early in embryogenesis. This was originally suggested by experiments with anti-E-cadherin antibodies [Hyafil et al., 1981], and recently also demonstrated genetically [Larue et al., 1994; Riethmacher et al., 1995]. A targeted mutation was introduced into the *E-cadherin* gene of the mouse, via homologous recombination and embryonic stem cell technology. This mutation removes sequences essential for Ca^{2+} binding and thus for the adhesive function of the molecule [Riethmacher et al., 1995]. Animals that carry this mutation in a heterozygous state appear normal and are fertile. However, homozygous mutant embryos do not develop normally. They can reach the morula stage and also compact properly, but the compacted state is not sustained. The individual morula cells in the embryo lose their morphological polarisation, they become rounded and continue to divide. As a consequence, the embryos appear totally distorted at a time when wild-type or heterozygous mutant embryos form a well-organised blastocyst with a well-formed blastocoele. Since the mutant embryos never emerge from the zona pellucida, further development, particularly implantation into the uterus, cannot take place. It has previously been reported that removal of Ca^{2+} ions or treatment with anti E-cadherin antibodies interferes with the compaction of the mouse morula, indicating that the initial compaction and polarisation of the epithelial-like cells is dependent on E-cadherin [Hyafil et al., 1981; Vestweber and Kemler, 1984]. Nevertheless, embryos which lack a functional E-cadherin gene can undergo compaction. This appear to be due to maternally derived E-cadherin, and not to functional compensation by other cell adhesion molecules. The maternal E-cadherin thus suffices for the initial compaction and for the formation of the first polarised, epithelial-like cells, but not for further development beyond the morula stage.

The integrity of the adherens junction also requires functional catenin molecules. Mutant cells that lack α-catenin have been described; the cadherin-mediated cell adhesion is not functional in such cells, and can be restored by the expression of *α-catenin* cDNA driven by an appropriate promoter [Hirano et al., 1992; Oda et al., 1992; Shimoyama et al., 1992]. Similar experiments with a cell line mutant for the *β-catenin* gene have demonstrated its importance for the function of the adherens junction [Hirohashi, pers. commun.]. By transient expression of cDNA encoding variant *α-* and *β-catenin* cDNAs which lack sequences for different subdomains, the interactions of the molecules in the adherence complex have been elucidated. E-cadherin binds directly to β-catenin; the binding site in β-catenin consists of armadillo repeat sequences (named after the *Drosophila armadillo* gene where this sequence motif was first identified) located in the central part of the molecule. The N-terminus of β-catenin interacts with α-catenin, and via α-catenin the cytoskeletal proteins bind to the complex. γ-Catenin (plakoglobin) is similar in structure to β-catenin, and it interacts in an analogous manner with E-cadherin and α-catenin [Huelsken et al., 1994]. Also the *APC* tumor suppresser gene product can bind to β-catenin. Mutations of

APC are frequently observed in colon carcinomas [Fearon et al., 1990]. APC and E-cadherin have indistinguishable binding sites on β-catenin, i.e. they can compete for β-catenin binding [Rubinfeld et al., 1993; Su et al., 1993; Naethke et al., 1994; Huelsken et al., 1994].

Epithelial-Mesenchymal Transitions Induced by Tyrosine Kinase Receptors

In addition, various soluble factors that function as ligands for tyrosine kinase receptors can induce transient EMT and motility in cell culture, like SF/HGF, aFGF, EGF and others [Jouanneau et al., 1993; Hoschuetzky et al., 1994]. The best characterised of these factors is the ligand for the c-met tyrosine kinase, that has been named SF [Théry and Stern, 1996] because of this activity on epithelial cells in culture [Stoker et al., 1987; Weidner et al., 1990]. In addition, the factor can induce growth of hepatocytes and therefore alternatively named HGF [Mijazawa et al., 1989; Nakamura et al., 1989]. A third activity of SF/HGF is the morphogenic action that was first discovered as the ability to induce epithelial cells to form tubules in vitro [Montesano et al., 1991a, b; Davies, 1996].

SF/HGF has a unique structure since it closely resembles proteases like plasminogen (40% sequence identity) but not other ligands for tyrosine kinase receptors [Miyazawa et al., 1989; Nakamura et al., 1989; Weidner et al., 1991]. The protein is produced as an inactive precursor molecule (90 kD) that is cleaved extracellularly into a heavy (60 kD) and a light (30 kD) chain [Hartmann et al., 1992; Lokker et al., 1992; Naldini et al., 1992]. The heavy chain contains an N-terminal hairpin loop and four kringle domains; the light chain shows extensive homologies to serine proteases. However, two of the three amino acids which form the catalytic triad of serine proteases are altered in SF/HGF, and therefore the factor has no catalytic activity. Further details on the kringle domain proteins are found in Théry and Stern [1996].

Like many other tyrosine kinase receptors, c-met was initially identified because of its transforming activity when mutated. The *met* oncogene was derived from an N-methyl-N-nitro-N-nitrosoguanidine-treated osteosarcoma cell line that was used in a transfection/tumorigenicity assay [Cooper et al., 1984]. The gene transferred from the osteosarcoma cells was the product of a rearrangement that fused *tpr* (translocated promoter region) on chromosome 1 to *c-met* on chromosome 7 [Park et al., 1986]. The oncogenic variant of *met* encodes a cytoplasmically located tyrosine kinase. In contrast, the protooncogene product is a transmembrane glycoprotein of 190 kD that is cleaved posttranslationally into an α- and a β-chain [Gonzatti-Haces et al., 1988; Giordano et al., 1989]. A major breakthrough in the understanding of the c-met receptor was the identification of its ligand, SF/HGF [Bottaro et al., 1991; Naldini et al., 1991]. All the known biological activities of SF/HGF are mediated by the c-met receptor [Weidner et al., 1993].

We have recently analysed the normal physiological roles of SF/HGF and c-met by introducing targeted mutations into the *SF/HGF* or the *c-met* genes of the mouse, via homologous recombination and embryonal stem cell technology [Schmidt et al., 1995; Bladt et al., 1995]. Whereas animals with heterozygous mutations in either the *SF/HGF* or the *c-met* gene are healthy and fertile, the homozygous mutations are not compatible with normal development. *SF/HGF* –/– or *c-met* –/– animals die between days 12.5 and 16.5 (E12.5 and E16.5) of development. The probable cause of the embryonal lethality is an impairment in development of the placenta. In particular, trophoblast cells are affected, and, consequently, a disorganisation of the labyrinth layer is observed [Vićovac and Aplin, 1996]. Moreover, the development of the liver is affected in mutant animals. The liver is considerably reduced in size and histological examination shows damage to the embryonic liver on E14.5 that varies in intensity and is not observed on E12.5. The essential function of *SF/HGF* and *c-met* in development of the liver is also supported by cell fate analysis of embryonic stem cells that carry two mutant alleles of *c-met:* Such cells cannot contribute to the liver (and to particular muscle groups, see also below), but participate in the development of other internal organs and cell types. Thus, two epithelial cell types in the embryo, hepatocytes and trophoblast cells of the placenta, are affected by the *SF/HGF* or *c-met* mutation. The c-met receptor is widely expressed on epithelial cells, both during development and in the adult, whereas SF/HGF is produced usually by mesenchymal cells in the vicinity [Sonnenberg et al., 1993]. SF/HGF and c-met therefore form a paracrine signalling system, a concept originally suggested by Stoker and colleagues, which is essential for normal development of the liver and placenta.

SF/HGF can also induce EMT in vivo. Dermomyotomal cells that are located at the ventrolateral edge of the somite undergo EMT, dissociate from the epithelial dermomyotome and consequently migrate as single cells to the anlagen of the limb, diaphragm and tongue. In *SF/HGF* –/– or *c-met* –/– animals, such migratory myogenic cells are not generated, and as a consequence, muscle in the limb, diaphragm and tongue do not form (fig. 2a, b) [Bladt et al., 1995]. Migratory myogenic precursor cells are only gener-

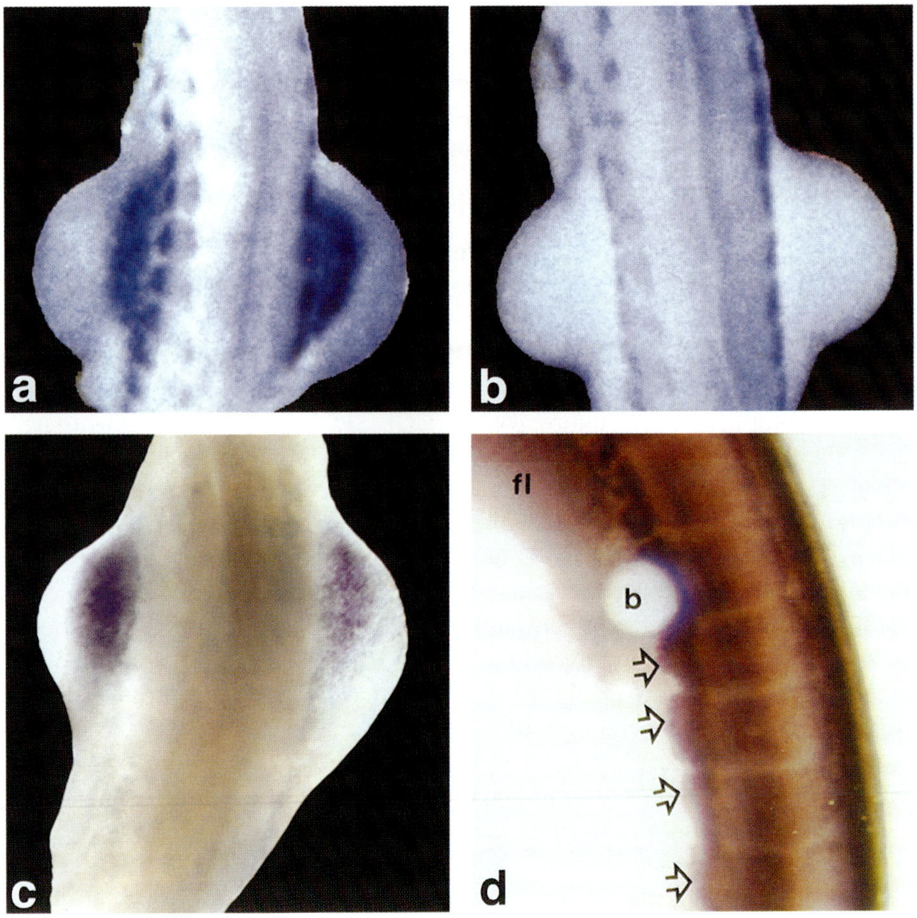

Fig. 2. SF/HGF induces EMT in vivo and generates migratory myogenic precursor cells. Myogenic precursor cells in wild-type embryos (**a**) and embryos homozygous for a mutation in c-*met* (**b**) as analysed by in situ hybridization with a c-*met*-specific probe. The precursor cells are present in somites of wild-type and mutant animals. In addition, migrating myogenic precursor cells are visualised in wild-type embryos (**a**), which emigrate from the somites and invade the anlage of the limb. In mutant embryos (**b**), such migrating cells are not observed. Note that the mutation introduced into the c-*met* gene removes sequences encoding the ATP-binding site of the c-met receptor, but leaves the transcription unit otherwise intact. Thus, homozygous mutant animals express met^{Δ} transcript which can be visualised in this experiment. **c** In vivo, the emigration is regulated by the temporarily and spatially controlled SF/HGF expression. The pattern of SF/HGF expression in the limb during migration of myogenic precursor cells as visualised by in situ hybridization. **d** Ectopic application of SF/HGF by application of a bead soaked in SF/HG in the chick embryo. The bead was implanted into the interlimb region of chick embryos and induces EMT and emigration of myogenic precursor cells, which are visualised by in situ hybridization with a *Pax-3*-specific probe. Under normal circumstances, interlimb somites never generate motile cells. The forelimb bud (fl) and the implanted bead (b) are indicated. Arrows point towards myogenic precursor cells that emigrate from the somite at the ectopic site.

ated from particular somites. Ectopic application of SF/HGF to somites at interlimb levels, which under normal conditions do not generate migratory cells, can induce EMT and emigration of cells (fig. 2d) [Brand-Saberi et al., 1996]. In vivo, this emigration from specific somites is regulated by precise temporal and spatial regulation of *SF/HGF* expression. Transcripts for *SF/HGF* are found in the limb buds (fig. 2c) and in branchial arches, and thus in the vicinity of somites that generate migratory myogenic precursor cells [Bladt and Birchmeier, unpubl. results]. Thus, SF/HGF controls EMT and the generation of motile cells in vivo. Morphologically, this normal developmental process is indistinguishable from the generation of motile cells as it is observed during metastasis.

Among the activities of SF/HGF that were described in cell culture, the morphogenic activity of SF/HGF is unique and no other factor with similar properties has been described [Montesano et al., 1991a, b]. MDCK cells, when grown in a three-dimensional collagen matrix for several days, form hollow cysts. When SF/HGF is added, individual cells dissociate and move away from the cysts. Subsequently, the cells reassociate and form continuous tubules. These tubules have a lumen surrounded by well-polarised epithelial cells with a smooth basal surface in contact with the collagen matrix, and an apical surface rich in microvilli that faces the lumen. The structures formed in vitro thus resemble the tubular epithelia present in many organs [Montesano et al., 1991a]. In the development of the mammary gland, SF/HGF plays a similar role, since it can induce the branching and growth of the tubular epithelia in organ cultures [Yang et al., 1995]. Recent experimental evidence indicates, however, that not only can the formation of tubular epithelia be induced by SF/HGF, but also other morphogenic programs such as formation of crypt-like structures from colon epithelial cells [Brinkmann et al., 1995]. Thus, SF/HGF seems able to activate the intrinsic morphogenic potential of various epithelial cell types, but does not instruct the cells to form one particular structure.

In vitro data indicate that SF/HGF does not only induce morphogenesis or motility of epithelial cell, but also invasiveness into collagen matrices [Weidner, 1990]. Furthermore, it has recently been shown that SF/HGF is produced by breast carcinoma cells, and that the level of expression correlates with the state of tumor progression [Yamashita et al., 1994]. How can the same factor be responsible for epithelial morphogenesis and invasiveness of epithelial cells? Interestingly, the different morphogenic programs can only be induced by SF/HGF in cells with intact epithelial characteristics, including expression of E-cadherin and functional adherens junctions [Brinkmann et al., 1995]. On the other hand, SF/HGF was shown to induce metastasis to the lymph nodes and lung when expressed by transfection of cDNA in dedifferentiated breast carcinoma cells [Meiners et al., unpubl. observations; Rosen et al., 1994]. Thus, the differentiation state of the epithelial cells seems to determine the response towards SF/HGF: morphogenesis requires an intact epithelial program, whereas metastatic behaviour is observed in cells that have already undergone EMT.

Epithelial-Mesenchymal Transitions in Malignancy

In human carcinomas, the loss of epithelial morphology is an important prognostic marker and correlates with a poor outcome of the disease. The loss of epithelial differentiation is particularly prominent at invasive fronts of the tumors, where the carcinoma cells break into the surrounding mesenchymal tissues. The invading cells can lose their epithelial appearance, become spindle-shaped and fibroblast-like, and, in the electron microscope, a reduced number of desmosomes are often observed [Weinstein et al., 1976; Gabbert et al., 1985]. Desmosomes (spot desmosomes) are the type of intercellular junctions most easily detected in intact tissues. Together with adherens junctions (belt desmosomes), they constitute the main adhering junctions of epithelial cells. Other known junctions do not primarily function in intercellular adhesion.

The prominent change in morphology of malignant carcinoma cells is associated with a loosening of intercellular adhesion. This is a consequence of a functional disturbance of cell-cell contacts. Apparently, the malignant cells downmodulate intercellular adhesiveness. That the loss of intercellular adhesion is a crucial step in carcinoma progression was first proposed 50 years ago [Coman, 1944; McCutcheon et al., 1948]. Disruption of the adherens junctions can be achieved in vitro by the addition of monoclonal antibodies against E-cadherin that interfere with the function of the cell adhesion molecule. The epithelial cells treated in this manner become fibroblast-like in appearance and invasive [Behrens et al., 1989]. An analysis of cell lines established from different types of carcinomas demonstrated that loss of epithelial morphology and intercellular adhesion often correlates with the loss or downregulation of *E-cadherin*. Carcinoma cells with a well-differentiated epithelial morphology do not invade a collagen matrix in vitro and produce E-cadherin, whereas carcinoma cells with a fibroblast-like morphology are motile, invade collagen matrices and have lost E-cadherin [Frixen et al., 1990]. Invasiveness of human carcinoma cells can be prevented by the transfection of *E-cadherin* cDNA under an ectopic promoter, and can be induced again by anti-E-cadherin antibodies. Similar results were obtained with normal murine mammary gland epithelia or MDCK cells transformed by v-*ras* or activated H-*ras*, respectively, which display a dissociated and invasive phenotype and then do not produce E-cadherin. Expression of *E-cadherin* cDNA under the control of an ectopic promoter in such cells restores cell adhesion and also abolishes the ability of cells to invade a collagen matrix or heart tissue [Vleminckx et al., 1991]. Therefore, invasive behaviour as assessed by various in vitro assays correlates with a fibroblast-like morphology and motility of carcinoma cells, and with the loss of E-cadherin-mediated cell adhesion.

Since the importance of intact E-cadherin-mediated cell adhesion in the maintenance of the epithelial phenotype and the prevention of invasion and metastasis was recognised [Behrens et al., 1989], various types of primary human carcinomas were examined for defects in the constituents of intercellular junctions. A correlation between dysfunction of E-cadherin-mediated cellular adhesion, by downregulation of *E-cadherin* and *α-catenin* expression or mutation in the *E-cadherin* gene, and invasive potential of many types of carcinomas has emerged from these studies. We will discuss selected examples, with the goal to particularly review the different mechanisms of alterations that have been observed.

Diffuse-type gastric carcinomas contain poorly adhesive cells that often express E-cadherin protein and mRNA. When the mRNA was analysed by reverse PCR, a high fre quency of altered *E-cadherin* transcripts were identified. 50% of all the tumors tested express shortened transcripts that lack exon 9 or 10 [Becker et al., 1994]; these exons encode part of the first Ca^{2+}-binding site of E-cadherin and thus sequences known to be essential for adhesive function. Analysis of the genomic *E-cadherin* DNA demonstrated that mutations are present in the corresponding splice junctions, and that the second, intact copy of the gene was frequently lost [Becker et al., 1994]. Mutations in one copy, and mutation or loss of the second intact allele are typical for genes that play a causative role in the development or progression of human tumors, i.e. for tumor suppressor genes.

A frequently observed alteration in malignant carcinoma cells of various tissues of origin is downregulation or even complete absence of E-cadherin protein. In head and neck cancers, for example, downregulation of E-cadherin expression correlates with loss of epithelial differentiation of the tumors and with the formation of lymph node metastasis [Schipper et al., 1991]. In breast carcinomas, a correlation between downregulation of E-cadherin expression and malignancy is observed for lobular, but not for ductal carcinomas [Moll et al., 1993; Gamallo et al., 1993]. In colorectal carcinomas, downregulation of E-cadherin was similarly observed to correlate with late stages of tumor progression [Dorudi et al., 1993; Kinsella et al., 1993]. In prostate and bladder carcinomas, downregulation of E-cadherin correlates with poor survival rates [Bringuir et al., 1994; Otto et al., 1994].

The molecular aspects of junctional deficiencies and the loss of differentiation of epithelial cells in carcinomas have turned out to be quite complex; many junctional proteins (E-cadherin, catenins) but also proteins of various signal cascades (src or tyrosine kinase receptors) can be involved [Geiger et al., 1995]. The critical proteins are affected by downregulation of expression, by mutation, or by phosphorylation. Often, downregulation of E-cadherin expression correlates with loss of tumor differentiation and progression to lymph node metastasis. The prognostic value of the measurement of E-cadherin expression will certainly be the scope of future clinical research. Also, several components of the epithelial differentiation program appear to be deregulated in concert in carcinomas, which indicates that epithelial master genes [Hay, 1995] are affected during tumorigenesis. Obviously, developmental studies aimed at the elucidation of mechanisms that regulate the epithelial phenotype will also contribute to our understanding of malignant aberrations of epithelial cells during carcinogenesis.

References

Aaronson, S. (1991) Growth factors and cancer. Science *254:* 1146–1153.

Becker, K.-F., M.J. Atkinson, U. Reich, I. Becker, H. Nekarda, J.R. Siewert, H. Hoefler (1994) E-Cadherin gene mutations provide clues to diffuse type gastric carcinomas. Cancer Res *54:* 3845–3852.

Beddington, R., J.C. Smith (1993) Control of vertebrate gastrulation: Inducing signals and responding genes. Curr Opin Genet Dev *3:* 655–661.

Behrens, J., M. Mareel, F. Van Roy, W. Birchmeier (1989) Dissecting tumor cell invasion: Epithelial cells acquire invasive properties after the loss of uvomorulin-mediated cell-cell adhesion. J Cell Biol *108:* 2435–2447.

Berndorff, D., R. Gessner, B. Kreft, N. Schnoy, A. Lajous-Petter, N. Loch, W. Reutter, M. Hortsch, R. Tauber (1994). Liver-intestine cadherin. Molecular cloning and characterization of a novel Ca^{2+}-dependent cell adhesion molecule expressed in liver and intestine. J Cell Biol *125:* 1353–1369.

Birchmeier, C., W. Birchmeier (1993) Molecular aspects of mesenchymal-epithelial interactions. Ann Rev Cell Biol *9:* 511–540.

Birchmeier, C., E. Sonnenberg, K.M. Weidner, B. Walter (1993) Tyrosine kinase receptors in the control of epithelial growth and morphogenesis during development. BioEssays *15:* 185–190.

Birchmeier, W., J. Behrens (1994) Cadherin expression in carcinomas: Role in the formation of cell junctions and the prevention of invasiveness. Biochim Biophys Acta *1198:* 11–26.

Bladt, F., D. Riethmacher, S. Isenmann, A. Aguzzi, C. Birchmeier (1995) Essential role for the c-met receptor in the migration of myogenic precursor cells into the limb bud. Nature *376:* 768–771.

Bottaro, D.P., J.S. Rubin, D.L. Faletto, A.M. Chan, T.E. Kmiecik, G.F. Vande Woude, S.A. Aaronson (1991) Identification of the hepatocyte growth factor receptor as the c-met protooncogene product. Science *251:* 802–804.

Brand-Saberi, B., T.S. Müller, J. Wilting, B. Christ, C. Birchmeier (1996) Scatter factor/hepatocyte growth factor (SF/HGF) induces emigration of myogenic cells at interlimb level in vivo. Dev Biol *179:* 303–308.

Bringuir, P., R. Umbas, H. Schaafsma, H. Karthans, F. Debruyne, J. Schalken (1994) Decreased E-cadherin immunoreactivity correlates with poor survival in patients with bladder tumors. Cancer Res *53:* 3241–3245.

Brinkmann, V., M. Sachs, M. Weidner, W. Birchmeier (1995) Scatter factor/hepatocyte growth factor induces tissue-specific morphogenic programs in epithelial cells. J Cell Biol *131:* 1573–1586.

Buxton, R., A. Magee (1992) Structure and interactions of desmosomal and other cadherins. Semin Cell Biol *3:* 157–167.

Coman, D.R. (1944) Decreased mutual adhesiveness, a property of cells from squamous cell carcinomas. Cancer Res *1:* 625–629.

Cooper, C.S., M. Park, D.G. Blair, M.A. Tainsky, K. Huebner, C.M. Croce, G.F. Vande Woude (1984) Molecular cloning of a new transforming gene from a chemically transformed human cell line. Nature *311:* 29–34.

Davies, J. (1996) Mesenchyme to epithelium transition during development of the mammalian kidney tubule. Acta Anat *156:*187–201.

Donalis, M., M. Cramer, M. Ringwald, A. Starzinsky-Powitz (1991) Expression of M-cadherin, a member of the cadherin multigene family, correlates with differentiation of skeletal-muscle cells. Proc Natl Acad Sci USA *88:* 8024–8028.

Dorudi, S., J.P. Sheffield, R. Poulsom, J.M. Northover, I.R. Hart (1993) E-cadherin expression in colorectal cancer: An immunocytochemical and in situ hybridization study. Am J Pathol *142:* 981–986.

Fearon, E.R., K.R. Cho, J.M. Nigro, S.E. Kern, J.W Simons, J.M. Ruppert, S.R. Hamilton, A. Preisinger, G. Thomas, K. Kinzler, B. Vogelstein (1990) Identification of a chromosome 18q gene that is altered in colorectal cancers. Science *247:* 49–56.

Frixen, U., J. Behrens, M. Sachs, G. Eberle, B. Voss, A. Warda, D. Loechner, W. Birchmeier (1991) E-cadherin mediated cell-cell adhesion prevents invasiveness of human carcinoma cell lines. J Cell Biol *111:* 173–185.

Fujimori, T., M. Takeichi (1993) Disruption of epitheloid cell-cell adhesion by exogenous expression of a mutated nonfunctional N-cadherin. Mol Biol Cell *4:* 37–47.

Gabbert, H., R. Wagner, R. Moll, C.D. Gerharz (1985) Tumor dedifferentiation: An important step in tumor invasion. Clin Exp Metast *3:* 257–279.

Gamallo, C., J. Palacios, A. Suarez, A. Pizarro, P. Navarro, M. Quintanilla, A. Cano (1993) Correlation of E-cadherin expression with differentiation grade and histological type in breast carcinoma. Am J Pathol *142:* 987–993.

Garratt, A.N., M.J. Humphries (1995) Recent insights into ligand binding, activation and signalling by integrin adhesion receptors. Acta Anat *154:* 34–45.

Garrod, D.R. (1993) Desmosomes and hemidesmosomes. Curr Opin Cell Biol *5:* 30–40.

Geiger, B., S. Yehuda-Levenberg, A.D. Bershadsky (1995) Molecular interactions in the submembrane plaque of cell-cell and cell-maturity adhesions. Acta Anat *154:* 46–62.

Giordano, S., M.F. Di Renzo, R.P. Narsimhau, C.S. Cooper, C. Rosa, P.M. Comoglio (1989) Biosynthesis of the protein encoded by the c-met proto-oncogene. Oncogene *4:* 1383–1388.

Gonzatti-Haces, M., A. Seth, M. Park, T. Copeland, S. Oroszlan, G.F. Vande Woude (1988) Characterization of the tpr-met oncogene p65 and the met proto-oncogene cell surface p140 tyrosine kinases. Proc Natl Acad USA *85:* 21–25.

Harland, R.M. (1994) Neural induction in *Xenopus.* Curr Opin Genet Dev *4:* 543–549.

Hartmann, G., L. Naldini, K.M. Weidner, M. Sachs, E. Vigna, P.M. Comoglio, W. Birchmeier (1992) A functional domain in the heavy chain of scatter factor/hepatocyte growth factor binds the *c-Met* receptor and induces cell dissociation but not mitogenesis. Proc Natl Acad Sci USA *89:* 11574–11578.

Hay, E.D. (1995) An overview of epithelio mesenchymal transformation. Acta Anat *154:* 8–20.

Hirano, S., N. Kimoto, Y. Shimyama, S. Hirohashi, M. Takeichi (1992) Identification of a neural α-catenin as a key regulator of cadherin function and multicellular organisation. Cell *70:* 293–301.

Hoschuetzky, H., H. Aberle, R. Kemler (1994) β-Catenin mediates the interaction of the cadherin-catenin complex with epidermal growth factor receptor. J Cell Biol *127:* 1375–1380.

Huelsken, J., W. Birchmeier, J. Behrens (1994) E-cadherin and APC compete for the interaction with β-catenin and the cytoskeleton. J Cell Biol *127:* 2061–2069.

Hyafil, F., C. Babinet, F. Jacob (1981) Cell-cell interaction in early embryogenesis: A molecular approach to the role of calcium. Cell *21:* 927–934.

Hynes, R.O. (1992) Integrins: Versatility, modulation, and signalling in cell adhesion. Cell *69:* 11–25.

Jouanneau, J., J. Gavrilowic, D. Caruelle, M. Jaye, G. Moens, J.-P. Caruelle, J.P. Thiery (1991) Secreted or non-secreted forms of acid fibroblast growth factor produced by transfected epithelial cells influence cell morphology, motility and invasive potential. Proc Natl Acad Sci USA *88:* 2893–2997.

Juliano, R.L., J.A. Varner (1993) Adhesion molecules in cancer: The role of integrins. Curr Opin Cell Biol *5:* 812–818.

Kemler, R. (1992) Classical cadherins. Semin Cell Biol *3:* 149–155.

Kemler, R. (1993) From cadherins to catenins: Cytoplasmic protein interactions and regulation of cell adhesion. Trends Genet *9:* 317–321.

Kinsella, A.R., B. Green, G.C. Lepts, C.L. Hill, G. Bowie, B.A. Taylor (1993) The role of the cell adhesion molecule E-cadherin in large bowel tumor cell invasion and metastasis. Br J Cancer *67:* 904–909.

Kintner, C. (1992) Regulation of embryonic cell adhesion by the cadherin cytoplasmic domain. Cell *69:* 225–236.

Kokai, Y., J.A. Cohen, J.A. Drebin, M.I. Greene, (1987) Stage- and tissue-specific expression of the *neu* oncogene in rat development. Proc Natl Acad Sci USA *84:* 8498–8501.

Larue, L., M. Ohsugi, J. Hirchenhain, R. Kemler (1994) E-cadherin null mutant embryos fail to form a trophoectoderm epithelium. Proc Natl Acad Sci USA *91:* 8263–8267.

Lokker, N.A., M.R. Mark, E.A. Luis, G.L. Bennet, K.A. Robbins, J.B. Baker, P.J. Godowski (1992) Structure-function analysis of heptocyte growth factor: Identification of variants that lack mitogenic activity yet retain high affinity receptor binding. EMBO J *11:* 2503–2510.

Mason, I., F. Fuller-Pace, R. Smith, C. Dickson (1994) FGF-7 (keratinocyte growth factor) expression during mouse development suggests roles in myogenesis, forebrain regionalization and epithelial-mesenchymal interactions. Mech Age Dev *45:* 15–30.

McCutcheon, M., D.R. Coman, F. Moore (1948) Studies on invasiveness of cancer. Cancer *1:* 460–467.

Miyazawa, K., H. Tsubouchi, D. Naka, K. Takahashi, M. Okigaki, N. Arakaki, H. Nakayama, S. Hirono, O. Sakiyama, K. Takashi, E. Ghoda, Y. Daikuhara, N. Kitamura (1989) Molecular cloning and sequence analysis of cDNA for human hepatocyte growth factor. Biochem Biophys Res Commun *163:* 967–973.

Moll, R., M. Mitze, U. Frixen, W. Birchmeier (1993) Differentiated loss of E-cadherin expression on infiltrating ductal and lobular breast carcinomas. Am J Pathol *143:* 1731–1742.

Montesano, R., G. Schaller, L. Orci (1991a) Induction of epithelial tubular morphogenesis in vitro by fibroblast-derived soluble factors. Cell *66:* 697–711.

Montesano, R., K. Matsumoto, T. Nakamura, L. Orci (1991b) Identification of a fibroblast-derived epithelial morphogen as hepatocyte growth factor. Cell *67:* 901–908.

Mori, S., T. Akiyama, Y. Yamada, Y. Morishita, I. Sugawara, K. Toyoshima, T. Yamamoto (1989) C-erbB2 gene product, a membrane protein commonly expressed on human fetal epithelial cells. Lab Invest *61:* 93–97.

Naethke, I., J. Hinck, J. Swedlow, J. Papkoff, W. Nelson (1994) Defining interactions and distributions of cadherin and catenin complexes in polarized epithelial cells. J Cell Biol *125:* 1341–1352.

Nakamura, T., T. Nishizawa, M. Hagiya, T. Seki, M. Shimonishi, A. Sugimura, K. Tashiro, S. Shimizu (1989) Molecular cloning and expression of human hepatocyte growth factor. Nature *342:* 440–443.

Naldini, L., E. Vigan, R. Nashiman, G. Gaudino, R. Zarnegar, G. Michalopoulos, P. Comoglio (1991) Hepatocyte growth factor (HGF) stimulates the tyrosine kinase activity of the receptor encoded by the protooncogene c-met. Oncogene *6:* 501–504.

Naldini, L., L. Tamagnone, E. Vigna, M. Sachs, G. Hartmann, W. Birchmeier, Y. Daikuhara, H. Tsubouchi, F. Blasi, P. Comoglio (1992) Extracellular proteolytic cleavage by urokinase is required for activation of hepatocyte growth factor. EMBO J *11:* 4825–4833.

Oda, T., Y. Kanai, Y. Shimoyama, A. Nagafuchi, S. Tsukita, S. Hirohashi (1993) Cloning of the human α-catenin cDNA and its aberrant mRNA in a human cancer cell line. Cancer Res *53:* 1696–1701.

Okazaki, M., S. Takeshita, S. Kawai, R. Kikuno, A. Tsujimura, A. Kudo, E. Amann (1994) Molecular cloning and characterization of OB-cadherin, a new member of the cadherin family expressed in polarized epithelial cells. J Biol Chem *269:* 12092–12098.

Orr-Urtreger, A., M. Bedford, T. Burakova, E. Arman, Y. Zimmer, A. Yayon, D. Givol, P. Lonai (1993) Developmental localization of the splicing alternatives of fibroblast growth factor receptor-2 (FGFR2). Dev Biol *158:* 475–486.

Otto, T., W. Birchmeier, U. Schmidt, K. Rembrink, J. Schipper, H. Ruebben, A. Raz (1994) Inverse relation of E-cadherin and autocrine motility factor receptor expression as prognostic factors in patients with bladder carcinoma. Cancer Res *54:* 3120–3123.

Pachnis, V., B. Mankoo, F. Constantini (1993) Expression of the c-ret protooncogene during mouse embryogenesis. Development *119:* 1005–1017.

Park, M., M. Dean, C. Cooper, M. Schmidt, S. O'Brian, D. Blair, G. Vande Woude (1986) Mechanism of met oncogene activation. Cell *25:* 895–904.

Peters, K., S. Werner, X. Liao, S. Wert, J. Whitsett, L. Williams (1994) Targeted expression of a dominant negative FGF receptor blocks branching morphogenesis and epithelial differentiation of the mouse lung. EMBO J *13:* 3296–3301.

Press, M., C. Cordon-Cardo, D. Slamon (1990) Expression of the Her2/neu protooncogene in normal human and fetal tissues. Oncogene *5:* 953–962.

Quirke, P., A. Pickles, N. Tuzi, O. Mohamdee, W. Gullick (1989) Pattern of expression of c-erbB2 oncoprotein in human fetuses. Br J Cancer *60:* 64–69.

Riethmacher, D., V. Brinkmann, C. Birchmeier (1995) A targeted mutation in the mouse E-cadherin gene results in defective preimplantation development. Proc Natl Acad Sci USA *92:* 855–859.

Rodriguez-Boulan, E., W. Nelson (1989) Morphogenesis of the polarized epithelial cell phenotype. Science *245:* 718–725.

Rosen, E., J. Knesel, I. Goldberg, L. Jin, M. Bhargava, A. Joseph, R. Zitnik, J. Wines, M. Kelley, S. Rockwell (1994) Scatter factor modulates the metastatic phenotype of the EMT6 mouse mammary tumor. Int J Cancer *57:* 706–714.

Rubinfeld, B., B. Souza, I. Albert, O. Mueller, S. Chamberlain, F. Masiarz, S. Munemitsu, P. Polakis (1993) Association of the APC gene product with β-catenin. Science *262:* 1731–1734.

Sano, K., H. Tamihara, R. Heimark, S. Obata, M. Davidson, T. St. John, S. Taketani, S. Suzuki (1993) Protocadherins: A large family of cadherin related molecules in central nervous system. EMBO J *12:* 2249–2256.

Schipper, J., U. Frixen, J. Behrens, A. Unger, K. Jahnke, W. Birchmeier (1991) E-cadherin expression in squamous cell carcinomas of head and neck: Invasive correlation with tumor dedifferentiation and lymph node metastasis. Cancer Res *51:* 6328–6337.

Schmidt, C., F. Bladt, S. Goedecke, V. Brinkmann, W. Zschiesche, M. Sharpe, H. Gherardi, C. Birchmeier (1995) Scatter factor hepatocyte growth factor is essential for liver development. Nature *373:* 699–702.

Schuchardt, A., V. D'Agati, L. Larsson-Blomberg, F. Constantini, V. Pachnis (1994) Defects in the kidney and enteric system of mice lacking the tyrosine kinase receptor ret. Nature *267:* 380–383.

Schwarz, M., K. Owaribe, J. Kartenbeck, W. Franke (1990) Desmosomes and hemidesmosomes: Constitutive molecular components. Ann Rev Cell Biol *6:* 461–491.

Shimoyama, Y., A. Nagafauchi, S. Fujita, M. Gotoh, M. Takeichi, S. Tsukita, S. Hirohashi (1992) Cadherin dysfunction in a human cancer cell line: Possible involvement of loss of α-catenin expression in reduced cell-cell adhesiveness. Cancer Res *52:* 1–5.

Simons, K., A. Wandinger-Ness (1990) Polarized sorting in epithelia. Cell *62:* 207–210.

Sonnenberg, A., J. Calafat, H. Janssen, H. Daams, L. van der Raaij-Helmer, R. Falcioni, S. Kennel, J. Aplin, J. Baker, M. Loizidou, D. Garrod (1991) Integrin α_6/β_4 complex is located in hemidesmosomes, suggesting a major role in epidermal cell-basement membrane adhesion. J Cell Biol *113:* 907–917.

Sonnenberg, E., A. Goedecke, B. Walter, F. Bladt, C. Birchmeier (1991) Transient and locally restricted expression of the rosl protooncogene during mouse development. EMBO J *10:* 3693–3702.

Sonnenberg, E., D. Meyer, K. Weidner, C. Birchmeier (1993) Scatter factor/hepatocyte growth factor and its receptor, the c-met tyrosine kinase, can mediate a signal exchange between mesenchyme and epithelia during mouse development. J Cell Biol *123:* 223–235.

Stoker, M., E. Gherardi, M. Perryman, J. Gray (1987) Scatter factor is a fibroblast-derived modulator of epithelial cell motility. Nature *327:* 239–242.

Su, L., B. Vogelstein, K. Kinzler (1993) Association of the APC tumor suppressor protein with catenins. Science *262:* 1734–1737.

Takeichi, M. (1991) Cadherin cell adhesion receptors as a morphogenic regulator. Science *252:* 1451–1455.

Théry, C., C. Stern (1996) Roles of Kringle domain-containing serine proteases in epithelial-mesenchymal transitions during embryonic development. Acta Anat *156:* 162–172.

Timpl, R. (1989) Structure and biological activity of basement membrane proteins. Eur J Biochem *180:* 487–502.

Troyanovsky, S., R. Troyanovsky, L. Eshking, R. Leube, W. Franke (1994) Identification of amino acid sequence motifs in desmocollin, a desmosomal glycoprotein, that are required for plakoglobin binding and plaque formation. Proc Natl Acad Sci USA *91:* 10790–10794.

Tsukita, S., M. Itoh, A. Nagafuchi, S. Yonemura, S. Tsugita (1993) Submembranous junctional plaque proteins include potential tumor supressor molecules. J Cell Biol *123:* 1049–1053.

Tsukita, S., S. Tsukita, A. Nagafauchi, S. Yonemura (1992) Molecular linkage between cadherins and actin filaments in cell-cell adherens junctions. Curr Opin Cell Biol *4:* 834–839.

Valles, A.M., B. Boyer, J.P. Thiery (1991) Adhesion systems in embryonic epithelial to mesenchymal transformation and in cancer invasion and metastasis; in Goldberg ID (ed): Cell Motility Factors. Basel, Birkhäuser, pp 17–34.

Vestweber, D., R. Kemler (1984) Rabbit antiserum against a purified surface glycoprotein decompacts mouse preimplantation embryos and reacts with specific adult tissues. Exp Cell Res *152:* 169–178.

Vićovac, L., J. Aplin (1996) Epithelial-mesenchymal transition during trophoblast differentiation. Acta Anat *156:* 202–216.

Viebahn, C. (1995) Epithelio-mesenchymal transformation during formation of the mesoderm in the mammalian embryo. Acta Anat *154:* 79–97.

Vleminckx, K., L. Vakaet, M. Mareel, W. Fier, F. Van Roy (1991) Genetic manipulation of E-cadherin expression by epithelial tumor cells reveals an invasion suppressor role. Cell *66:* 107–119.

Weidner, K.M., J. Behrens, J. Vandekerckhove, W. Birchmeier (1990) Scatter factor: molecular characteristics and effect on the invasiveness of epithelial cells. J Cell Biol *111:* 2097–2108.

Weidner, K.M., N. Arakaki, G. Hartmann, J. Vandekerckhove, S. Weingart, H. Rieder, C. Fonatsch, H. Tsubushi, T. Hishida, Y. Daikuhara, W. Birchmeier (1991) Evidence for the identity of human scatter factor and human hepatocyte growth factor. Proc Natl Acad Sci USA *88:* 7001–7005.

Weidner, K.M., M. Sachs, W. Birchmeier (1993) The *Met* receptor tyrosine kinase transduces motility, proliferation and morphogenic signal of scatter factor/hepatocyte growth factor in epithelial cells. J Cell Biol *121:* 145–154.

Weinstein, R.S., F.B. Merck, J. Alroy (1976) The structure and function of intercellular junctions in cancer. Adv Cancer Res *23:* 23–89.

Werner, S., K. Peters, M. Longaker, F. Fuller-Pace, M. Banda, L. Williams (1992) Large induction of keratinocyte growth factor expression in the dermis during wound healing. Proc Natl Acad Sci USA *89:* 6896–6900.

Werner, S., W. Weinberg, X. Liao, K. Peters, M. Blessing, S. Yuspa, R. Weiner, L. Williams (1993) Targeted expression of a dominant negative FGF receptor mutant in the epidermis of transgenic mice reveals a role of FGF in keratinocyte organization and differentiation. EMBO J *12:* 2635–2643.

Werner, S., H. Smola, X. Liao, M. Longaker, T. Krieg, P. Hofschneider, L. Williams (1994) The function of KGF in morphogenesis of epithelium and reepithelization of wounds. Science *266:* 819–822.

Yamashita, J., M. Ogawa, S. Yamashita, K. Nomura, M. Kuramoto, T. Saishoij, S. Shin (1994) Immunoreactive hepatocyte growth factor is a strong and independent predictor of recurrence and survival in human breast cancer. Cancer Res *54:* 1630–1633.

Yang, J., E. Spitzer, D. Meyer, M. Sachs, C. Niemann, G. Hartmann, K.M. Weidner, C. Birchmeier, W. Birchmeier (1995) Sequential requirement of scatter factor/hepatocyte growth factor and neu differentiation factor/heregulin in the morphogenesis and differentiation of the mammary gland. J Cell Biol *131:* 215–226.

Acta Anatomica

Acta Anat 1996;156:227–239

B. Boyer
A.M. Vallés
J.P. Thiery

Laboratoire de Dynamique et
Compartimentation Cellulaires,
UMR CNRS 144,
Institut Curie Section de Recherche,
Paris, France

Model Systems of Epithelium-Mesenchyme Transitions

Key Words
Cell scattering
Growth factors
Extracellular matrix
Signaling pathway
Epithelial-mesenchymal transitions

Abstract
This contribution discusses which factors bring about the transformation of epithelium to mesenchyme. Amongst biological agents which have this role are a number of cytokines (e.g. EGF, FGF-1, TGF-β, HGF/SF) and extracellular matrix macromolecules, such as collagens. The coordinated changes in cell morphology, associated with the induction of cell motility and the loss of intercellular junctions, are under the control of signaling molecules that transduce the signal emanating from the plasma membrane, which ultimately lead to changes in gene expression.

Introduction

Epithelia are tissues assuming specialized and specific functions. Simple epithelia bordering the lumen in the gastrointestinal tract or renal tubules, for example, are involved in the vectorial transport of soluble molecules while stratified epithelia, such as epidermis, are implicated in the mechanical resistance to external stresses. Once established as epithelial, cells are not obligatorily maintained in this state of differentiation. Under specific circumstances, epithelial cells can modify their program of differentiation and escape from epithelial structures. In that case, they are converted into mesenchymal or fibroblast-like cells, capable of migrating freely throughout the connective tissue (see fig. 1 for a schematic drawing of EMT). This process, termed epithelium mesenchyme transition (EMT) [Hay, 1995], participates in morphogenetic remodellings in embryonic life: examples of EMTs include the formation of mesoderm in the sea urchin embryo, the migration of neural crest cells (NCC) out of the neural tube [Duband et al., 1995], the formation of the cardiac valves [Markwald et al., this vol.] and the formation of midline mesenchyme during palatal fusion. EMTs may also be involved in physiological or pathological processes of adult life, where transient or permanent transformations of cells of epithelial origin into fibroblastic-like motile cells occur. In the context of the etiopathology of cancer, it is of fundamental importance to determine the exact role played by EMT-like modifications in the dispersion of carcinoma cells, eventually leading to cancer cell metastasis [Birchmeier et al., this vol.]. In this respect, it is noteworthy that epithelial cell dispersion has been described to occur via processes referred to as motogenic. The relationship between motogenic processes and EMT remains elusive, but there is no doubt that motogenic processes, like EMT, require the breakdown of intercellular junctions, and extended changes in cell architecture to allow cell movement. We can therefore presume that the 'motogenicity' of epithelial cells is not achieved unless EMT-like changes have taken place.

Abbreviations used in this paper:
ECM = Extracellular matrix; EGF = epidermal growth factor; EMT = epithelium-mesenchyme transition; FGF = fibroblast growth factor; NCC = neural crest cell; TGF = transforming growth factor; SF/HGF = scatter factor/hepatocyte growth factor.

KARGER

E-Mail karger@karger.ch
Fax + 41 61 306 12 34
http://www.karger.ch

Dr. Brigitte Boyer
Laboratoire de Dynamique et
Compartimentation Cellulaires
UMR CNRS 144
Institut Curie Section de Recherche
26, rue d'Ulm, F–75231 Paris Cedex 05 (France)

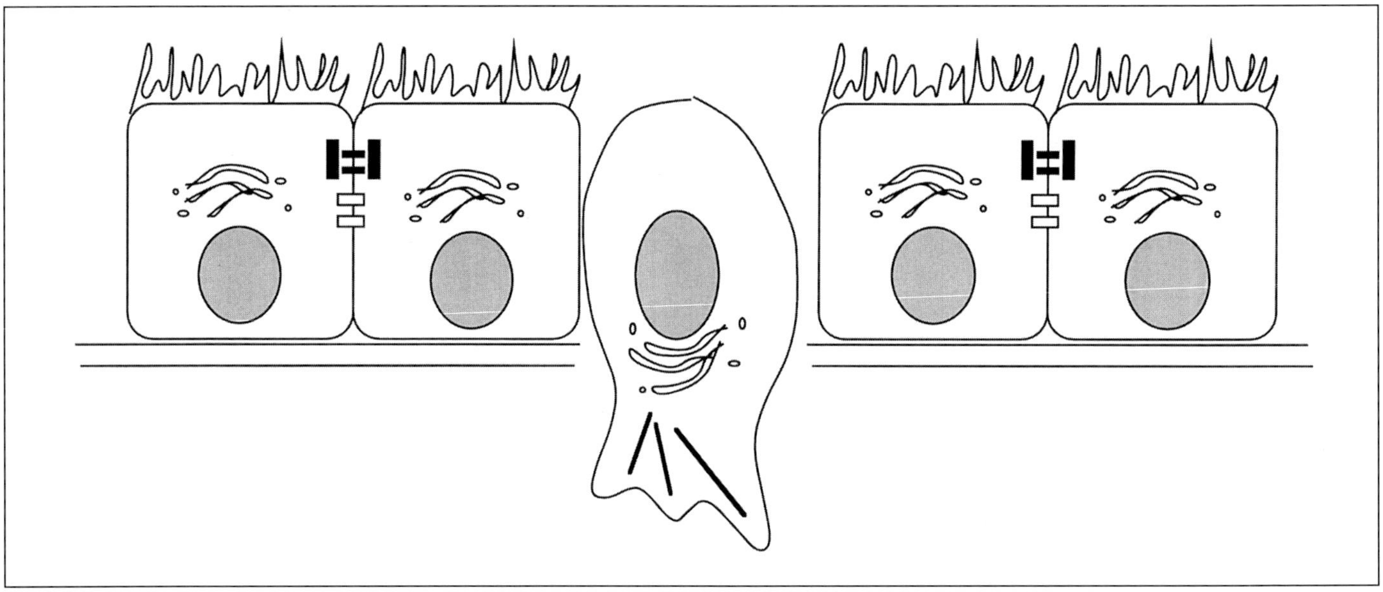

Fig. 1. Schematic representation of the EMT process. In response to extrinsic or intrinsic signals, epithelial cells turn off genes encoding junction-associated proteins, start to express vimentin and mesenchymal front end-back end polarity that allows them to elongate and move out through gaps in the basement membrane.

Is EMT an Inductive Event?

Considering that, in embryo, EMTs arise in precise locations and times, the question arises as to whether EMTs are 'cell-autonomous' processes that are programmed by a cellular internal clock or whether they are induced by external factors to which cells are sensitized. The question has been largely debated in the case of the migration of NCC. The fact that explants of neural tubes are spontaneously capable of giving rise to migrating NCC suggests that the induction of EMT can proceed under the control of an internal program. However, these experiments cannot rule out the possibility that inducers of EMT could be secreted by cells located in the vicinity of NCC. Accordingly, transforming growth factor (TGF)-β has been shown to accelerate the departure of NCC from in vitro explanted neural tubes [Duband et al., 1995], and also mediates in vitro the EMT of endothelial cells of the atroventricular canal into progenitor cells of the valves and septa in the embryonic heart [Potts and Runyan, 1989; Markwald et al., this vol.]. However, these experiments have not demonstrated the requirement of TGF-β in the induction of EMT processes.

In vitro studies of cell lines established from normal epithelial tissues also reported on the spontaneous generation of fibroblast-like cells developing in the epithelial monolayer [Dulbecco et al., 1981; Benali et al., 1991]. However, the conditions of culture, and particularly the presence of fetal calf serum in the culture medium, appeared to be a determinant in the induction of EMT [Benali et al., 1991]. It therefore remains unclear whether or not the apparently spontaneous transition was in fact dependent on soluble factors contained in the medium. These examples clearly show the need of in vitro models of EMT, but also demonstrate that in vivo studies are necessary to fully demonstrate the role of factors shown in vitro to be potential inducers.

The in vitro studies devoted to the analysis of EMT have led to the description of different modes of induction. Interestingly, there is a network of potential interplays between the distinct types of inducers that could explain how EMT is regulated. These topics will be discussed here. It should be noted that in vitro-induced EMTs are generally reversible. Therefore, if EMT can be regarded, in many aspects, as a transdifferentiation process between epithelial and mesenchymal programs, it does not proceed by irreversible steps, as authentic differentiation events usually do.

The Different Modes of EMT Induction

Growth Factors

There is increasing evidence that growth factors, discovered on the basis of their growth-promoting effect, are multipotential, eliciting a wide range of biological responses. Interestingly, their effect on cell differentiation appears to

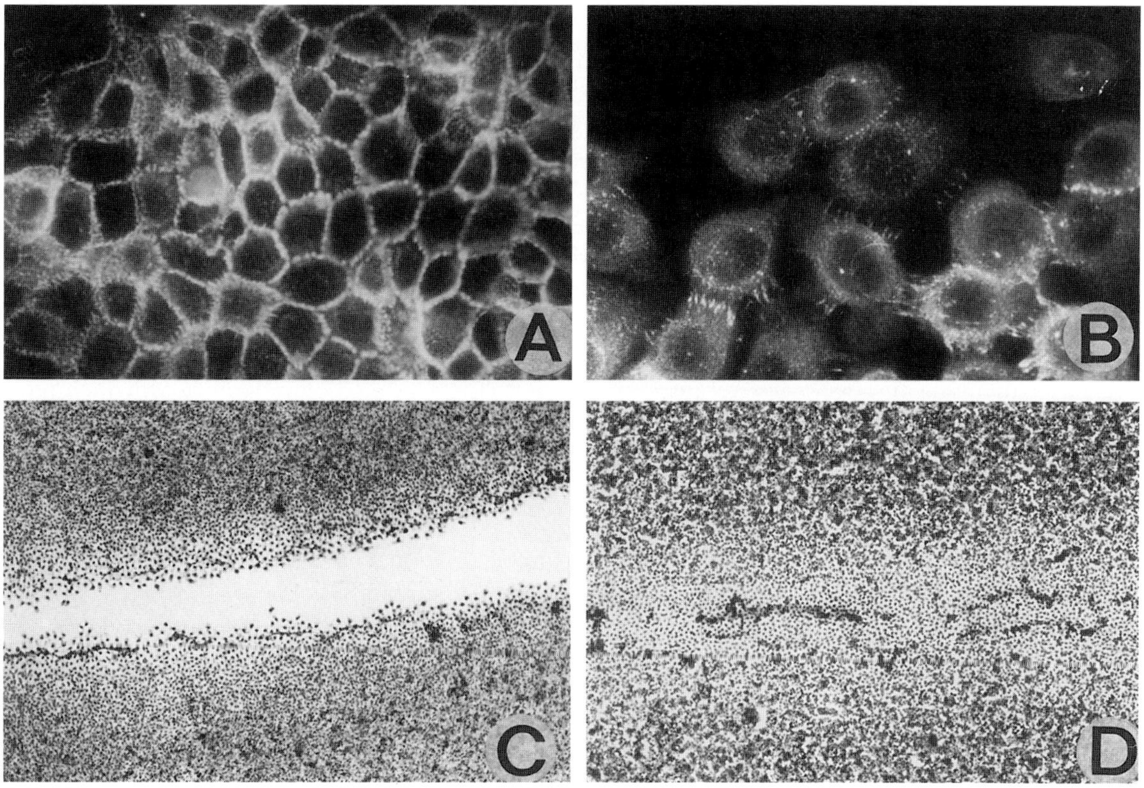

Fig. 2. FGF-1-induced EMT of NBT-II cells is accompanied by cell dissociation and acquisition of cell motility. The NBT-II rat bladder carcinoma cells were processed for immunofluorescence microscopy with a desmoplakin monoclonal antibody (**A**, **B**). Alternatively, NBT-II cell monolayers were scratched in order to produce a wounded area. Wound repair resulted from the migration of cells into the gap and therefore reflected cell motility. NBT-II cells were cultured under standard conditions (**A**, **C**) or in the presence of 30 ng/ml FGF 1 (**B**, **D**) for 18 h. In (**A**), cells are linked by numerous desmosomes, as inferred from the bright desmoplakin cortical staining. After incubation with FGF-1 (**B**), desmoplakin disappears from the cell surface and is concentrated in the cytoplasm, suggesting that desmosomes are internalized. In the 'wound healing assay' done in the absence of FGF-1 (**C**), NBT-II cells are not able to cover the wounded area, while they totally repair the gap in the presence of 30 ng/ml FGF-1 (**D**).

involve requirements opposed to those implicated in mitogenesis, since it is well established that cell proliferation and differentiation are antagonistic processes: cells usually cease dividing when they differentiate and, conversely, differentiation events are inhibited when cells enter the cell cycle. The EMT-inducing effect of growth factors can be related to a differentiating activity, in the sense that, in general, the growth factors cannot elicit simultaneously mitogenesis and EMT responses [Shiota et al., 1992; Vallés et al., 1990b; Geimer and Bade, 1991; Miettinen et al., 1994]. Accordingly, the PDGF-stimulated motogenic response of fibroblasts is maximal in growth-arrested cultures [Grotendorst, 1984; Hughes and McCulloch, 1991]. However, there are exceptions to this rule. Epitaxin, a fibroblast-derived motility factor for epithelial cells simultaneously increases cell proliferation [Shimonaka and Yamaguchi, 1994].

Many growth factors have been demonstrated to promote in vitro EMTs. EGF promotes the in vitro migration of normal keratinocytes [Barrandon and Green, 1987], intestinal epithelial cells [Blay and Brown, 1985], liver epithelial cells [Bade and Feindler, 1988] and of pancreatic carcinoma cells [Klemke et al., 1994], and induce the generation of fibroblastic-like cells from normal mammary epithelial cells in culture [Stampfer and Bartley, 1988]. TGF-α, which binds to the cellular receptors to EGF converts bladder carcinoma cells into fibroblasts [Gavrilovic et al., 1990]. Insulin-like growth factor (IGF) stimulates the migration of rhabdomyosarcoma and melanoma cells [El Badry et al., 1990; Stracke et al., 1988]. We have found that fibroblast growth factor (FGF)-1 elicits a scattering response from NBT-II bladder carcinoma cell line [Vallés et al., 1990a]. Figure 2 shows the changes in cell-cell adhesion and motility of NBT-II cells after addition of FGF-1: desmosomes disappear from the cell periphery, as inferred from immunofluorescence labeling with anti-desmoplakin antibodies. Simultaneously, cells acquire the ability to move, as deduced from their capacity to repair a 'wound' made in the cell culture.

The prototype of growth factors endowed with a scattering activity is hepatocyte growth factor (HGF), also named scatter factor (SF) [Théry and Stern, this vol.], secreted by mesenchymal cells and capable of dispersing a variety of normal and tumor epithelial cells [Rosen et al., 1994]. However, a challenging discovery has recently been reported. When mouse fibroblasts are transfected with cDNAs encoding human SF and its cognate receptor (c-met), they are converted into epithelial-like cells [Tsarfaty et al., 1994]. Accordingly, anti-SF antibodies inhibit the differentiation of metanephric mesenchymal cells into epithelial precursors of metanephric rudiments grown in culture [Woolf et al., 1995]. If the positive effects of SF on two opposite processes (i.e. EMT and MET) are to be reconciled, it implies that SF would act as an 'interconverting factor' between epithelial and mesenchymal programs, suggesting that the two apparently antagonistic transdifferentiation events could share common control steps. Elucidating the precise contribution of growth/scatter factors during these morphogenetic events therefore awaits further characterization.

Extracellular Matrix (ECM)

As for growth factors, the effects of matrix macromolecules on cell motility is well documented. The requirement of cell adhesion to ECM in the generation of cell movement is absolute, since cells migrate by exerting tractional forces on their substrate. However, many observations have pointed to the fact that cell-matrix interactions established during cell locomotion must be labile, in order to allow cell translocation. Requirements of cell-matrix interactions have been demonstrated in many types of migratory cells: for example, neural crest cells utilize ECM fibronectin and tenascin for their oriented emigration from the neural tube [Erickson, 1990], and the migration of retinal ganglion axons on astroglial cells is mediated in part by laminin [Neugebauer et al., 1988]. However, the requirement of ECM molecules in promoting and sustaining cell movements is not necessarily associated with a role for these matrix components in the induction of EMT. A detailed survey of the literature dealing with cell motility per se is beyond the scope of the present review, which will focus on the promotion and modulation of EMT events by matrix macromolecules.

As in the case of growth factors, there is no doubt that the ECM can exert differentiating activities [Hay, 1993]. A well-studied example is provided by the mouse mammary gland in vitro system, demonstrating the requirement of basement membrane for the establishment and maintenance of a differentiated phenotype of the mammary epithelial cells [Lin and Bissell, 1993]. Remarkably, transdifferentiation from an epithelial program of differentiation to a mesenchymal one, which defines EMT events, is switched on by culturing a variety of differentiated epithelial cells in type I collagen gels [Hay, 1995]. This EMT event can be inhibited by retinoic acid, which promotes basement membrane formation, suggesting that the basal lamina, which determines the maintenance of an epithelial phenotype, can counteract the EMT-promoting effects of collagen. We also demonstrated that collagen fibrils, and not only tridimensional collagen gels, can induce EMT of carcinoma cells in culture [Tucker et al., 1990].

We and others reported on the modulation of the scattering activity of growth factors by specific matrix molecules: SF-induced scattering of MDCK cells is augmented in the presence of fibronectin [Clark, 1994], while FGF-1-promoted dispersion of NBT-II carcinoma cells is enhanced in the presence of collagen [Tucker et al., 1992]. These observations raise the question as to whether the matrix-mediated cellular signals leading to EMT interact positively with those induced by growth factors or whether ECM elements display positive effects on cell dispersion by acting mechanically as permissive substrates for cell locomotion.

On the other hand, matrix molecules can repress the cytokine-promoted EMT, as fibronectin can inhibit the EGF-induced scattering of liver epithelial cells [Bade and Nitzgen, 1985]. Whether they exert their action by interfering with EGF-mediated signals or by simply acting as nonpermissive substrates for cell translocation remains unclear.

Downstream Targets and Signals in Growth Factor and ECM-Mediated EMTs

Cell Adhesion Molecules

Adhesion molecules are obvious candidates for mediating the scattering effects of growth factors and ECM. Epithelial cells are indeed connected by intercellular junctions in which cell adhesion molecules (CAMs) assume the recognition and adhesion roles [Prieto and Crossin, 1995; Geiger et al., 1995] (fig. 3 for a schematic drawing of adherent junctions and desmosomes). During EMT, cell dissociation occurs, indicating that intercellular adhesion is abrogated. In the NBT-II carcinoma cell line, FGF-1 induces the

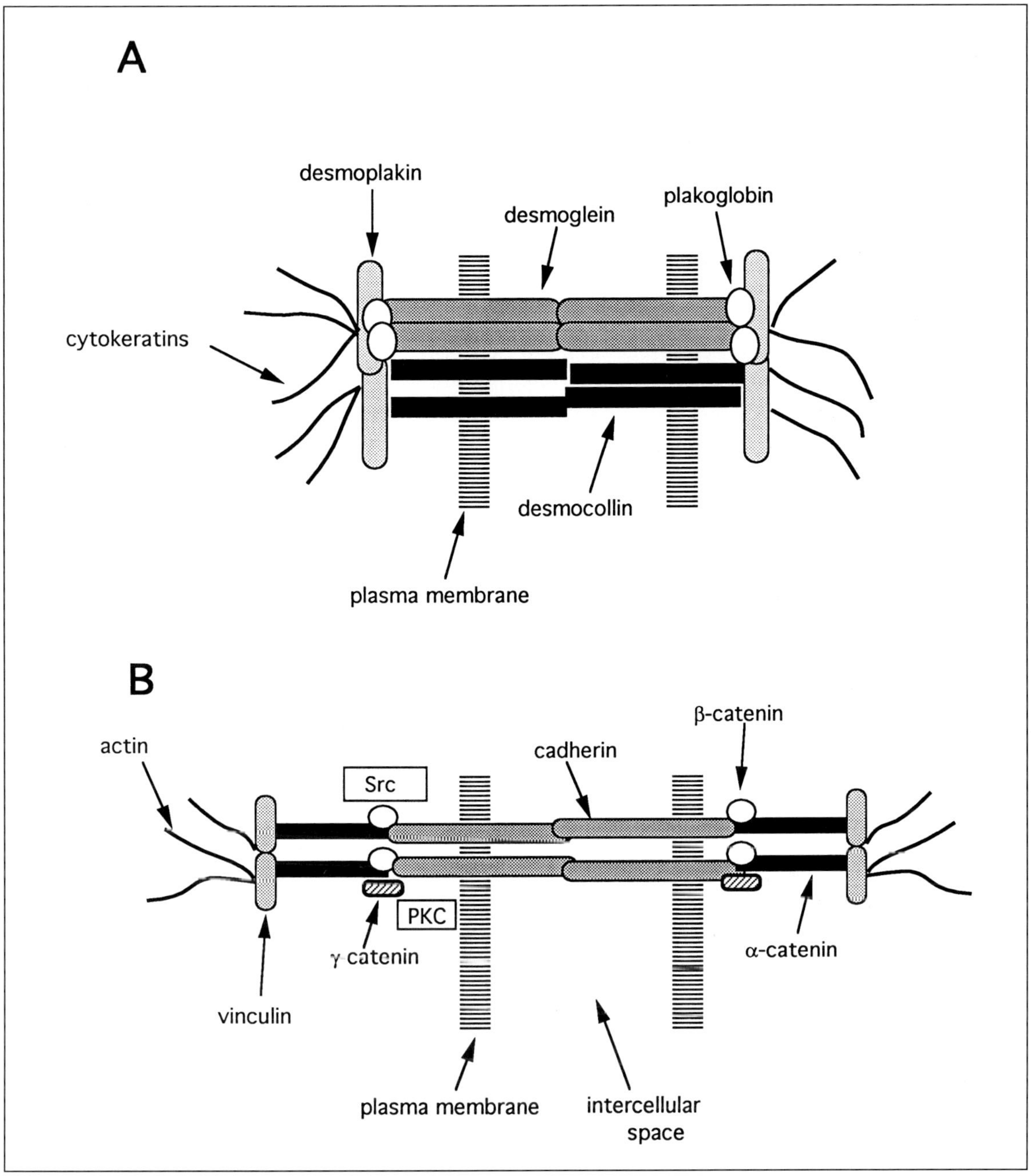

Fig. 3. Schematic representation of epithelial junctions. **A** Desmosome. **B** Adherent junction. Only the major protein interactions are represented. The illustrated location of some signaling molecules is tentative.

breakdown and internalization of desmosomes, correlating with cell dispersion. Among the different components of desmosomes, desmoglein, a cadherin-like cell-adhesion molecule is rapidly downregulated [Boyer et al., 1989].

Much attention has been paid to E-cadherin modulation during EMT processes, since it was long recognized that this calcium-dependent cell adhesion molecule plays a fundamental role in the establishment and maintenance of epithelial structures [Mege et al., 1988; McNeill et al., 1990]. In agreement with the prediction, E-cadherin is lost from colon carcinoma cells undergoing a phorbol-ester-induced cell scattering [Fabre and de Herreros, 1993]. Accordingly, a partial or complete absence of E-cadherin expression has been observed in several types of carcinomas of high invasive potential associated with a dedifferentiated phenotype [Mayer et al., 1993; Bringuier et al., 1993; Oka et al., 1993; Gamallo et al., 1993; Birchmeier et al., this vol.]. However, cell scattering, correlating with an increase in invasive and metastatic potentials, is not necessarily associated with a downmodulation of E-cadherin, although the adhesive properties of the carcinoma cells are clearly altered.

Cadherin function depends on its interactions with cytoplasmic molecules, the catenins, which mediate the binding of the C-terminal portion of cadherins with actin filaments [Gumbiner, 1993]. In some cases, alteration in cell-cell adhesion in carcinoma cells results from the loss of α- or β-catenin, due to gene mutations [Shimoyama et al., 1992; Kawanishi et al., 1995]. In other cases, no downmodulation of any component of the cadherin-adhesive complex was observed. During the FGF-1-induced EMT of NBT-II carcinoma cells, E-cadherin expression is not modified. Furthermore, transfection of E-cadherin cDNA into NBT-II cells, leading to stable overexpression of the molecule, does not prevent cells from FGF-1-induced dissociation and scattering, indicating that EMT does not necessarily result from a downregulation of E-cadherin [Boyer et al., 1991]. On the other hand, posttranscriptional modifications of the cadherin adhesion system [Geiger et al., 1995] might account for the reduced cell-cell adhesion observed during in vitro EMT or in vivo scattering of carcinoma cells. Shibamoto and colleagues found β-catenin and plakoglobin/γ-catenin tyrosine phosphorylated during EGF and SF-stimulation of in vitro carcinoma cell scattering [Shibamoto et al., 1994]. Concurrently, $pp60^{v\text{-}src}$ was shown to cause tyrosine phosphorylation of E-cadherin and β-catenin [Behrens et al., 1993], of β-catenin alone [Matsuyoshi et al., 1992] or of α- and β-catenins [Hamaguchi et al., 1993], associated with inactivation of the cadherin adhesion system. However, there was in no case a disruption of the cadherin-catenin complex as a result of tyrosine phosphorylation. The effects of catenin phosphorylation on the function of the adhesive complex remain therefore to be clarified.

Tyrosine kinases of the Src family phosphorylate other substrates related to β-catenin: p120, previously implicated in EGF-, PDGF- and CSF-1-induced signaling has been shown to contain 4 copies of the Arm repeat found in β-catenin, plakoglobin, and *armadillo*, the *Drosophila* homolog of plakoglobin [Reynolds et al., 1992]. p120 is associated with E-cadherin complexes and, like β-catenin and plakoglobin, its Src-induced tyrosine phosphorylation does not disrupt its association with the cadherin-based adhesion complex [Reynolds et al., 1994].

Interestingly, β-catenin associates directly with the tumor suppressor protein adenomatous polyposis coli (APC), which is frequently mutated in colocarcinomas [Rubinfeld et al., 1993; Su et al., 1993]. Association with APC occurs via the Arm domain of β-catenin, suggesting that Arm could play an important role for mediating protein-protein interactions in adhesion complexes. Therefore, an important question would be to determine whether tyrosine phosphorylation of the catenin family, mediated by scatter factors and oncogenic Src kinase, affects the association between APC and catenins and how this could lead to alterations in cell-cell adhesion.

The cadherin complex is also subjected to positive regulations, leading to increased adhesiveness. In this respect, several reports have provided evidence that the signaling molecule *Wnt-1* upregulates the expression of E-cadherin and plakoglobin [Bradley et al., 1993], stabilizes the association between β-catenin and E-cadherin [Hinck et al., 1994], and induces an epithelioid transformation of PC12 cells [Shackelford et al., 1993]. Understanding how negative and positive regulations of E-cadherin complexes are orchestrated during EMT events is a major goal in the field.

Integrins

Matrix macromolecules can trigger in vitro EMTs and are also essential for sustaining the migratory behavior of mesenchymal cells generated by EMT processes. For these reasons, integrins, the cellular receptors for ECM components, are most likely involved in EMT. Each integrin is composed of an α- and a β-subunit linked by noncovalent bonds. The 13 different α- and 8 β-subunits identified to date can combine to form the 21 known integrins with different binding specificities. The large family of cell-surface integrins mediate cell-cell and cell-matrix interactions [Gumbiner, 1993]. In fibroblasts, integrins are mainly found at the basal cell surface, where they often cluster in complexes called focal adhesions or focal contacts [Burridge et al., 1988] (fig. 4). At first, integrins were considered as mechanical links between matrix and cells. It was then recognized that integrins participate in signaling processes [Geiger et al., 1995], eventually leading to cell remodelling [Hynes, 1992]. On the other hand, integrins have to be activated in order to exert their adhesive function [Garratt and Humphries, 1995]. The molecular steps that couple cell activation to a change in the ability of an integrin to bind its extracellular ligand are far from being elucidated. However, changes in integrin adhesiveness probably arise from modifications in the affinity of the integrins for their ligands. The effector mechanisms that mediate this change in integrin affinity are not understood; they might include phos-

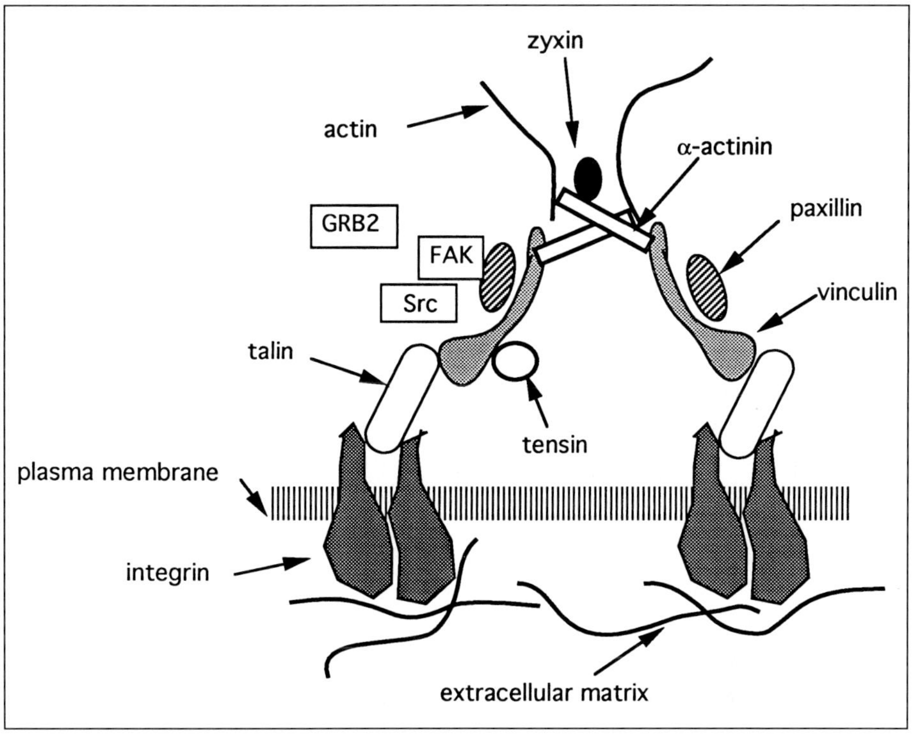

Fig. 4. Model for focal contact structure. The integrins bind to matrix macromolecules and at the same time interact with cytoskeletal components, which connect the transmembrane integrins with the actin filaments. The illustrated location of signaling molecules is tentative.

phorylation of the cytoplasmic tail of the α- and β-subunits [Diamond and Springer, 1994].

Among matrix macromolecules, collagen is the major component capable of inducing in vitro EMT [Hay, 1995]. As we demonstrated, while collagen and FGF-1 can induce separately the EMT of NBT-II carcinoma cells, the combination of both inducers has a synergistic effect on the scattering response, suggesting that common denominators exist in the two inducing machineries [Tucker et al., 1991]. Integrins, and particularly the collagen-specific integrins are obvious candidates for playing this role. In that context, it is interesting to note that the EGF-induced EMT of a normal mammary epithelial cell line can be specifically inhibited by blocking antibodies directed against the integrin α2β1, which is a major collagen-binding integrin [Matthay et al., 1993]. α2β1 might therefore represent one key element in the regulation of matrix and growth factor-induced EMT. α2β1 is also implicated in collagen-induced morphogenesis of normal mammary epithelial cells [Berdichevsky et al., 1992], and in collagen gel contraction [Schiro et al., 1991]. It is also implicated in metastasis of rhabdomyosarcoma cells [Chan et al., 1991] and is a marker of melanoma cancer progression [Klein et al., 1991]. However, its expression is often reduced in carcinomas [Juliano and Varner, 1993]. Therefore, the role of α2β1 in pathological EMT processes leading to carcinoma cell invasion and metastasis remains uncertain. How the signal caused by integrin ligation to collagen can produce the same effects as those generated by growth factor binding to its cellular receptors, and how the two types of inductions can co-operate are unsolved questions. However, signaling molecules involved in transducing the message emanating from activated growth factor receptors have recently been shown to participate in the integrin-mediated signaling pathway and could therefore represent converging points in the growth factor and ECM-induced EMTs (see below).

Signaling Pathways

With respect to growth factor-induced EMT, two related questions could be formulated:

(1) Is EMT regulated by a transduction cascade, initiated at the plasma membrane by the ligation of growth factor receptors and completed in the nucleus by the specific regulation of mRNA transcription?

(2) What is the molecular basis for the cellular decision allowing cells to escape from growth factor-induced mitogenesis and to enter a pathway leading to EMT?

As a first approach, we have observed that confluent cultures of NBT-II carcinoma cells are stimulated to enter mitosis but not EMT, and, conversely, that sparse cultures undergo EMT but are growth-inhibited upon stimulation with FGF-1 [Vallés et al., 1990b]. The culture conditions in

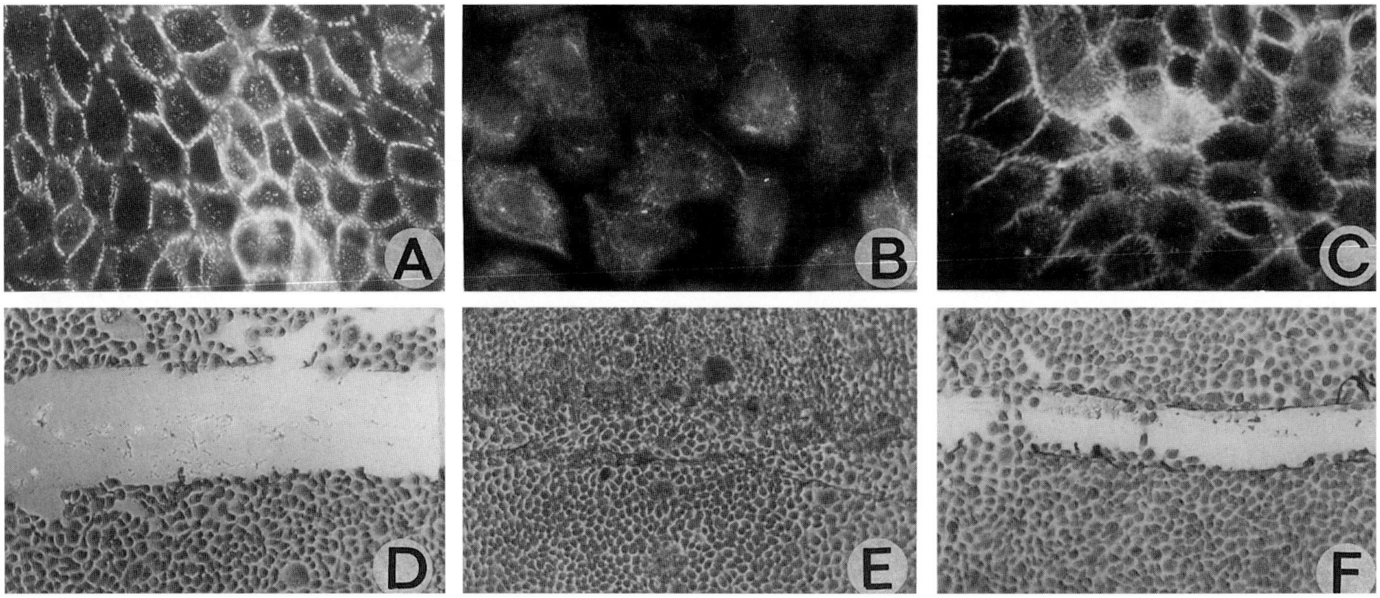

Fig. 5. The FGF-1-induced EMT of NBT-II cells is under control of a signaling pathway that is inhibited by raising the concentration of intracellular cAMP. **A–C** Immunofluorescence labeling of NBT-II cells with desmoplakin monoclonal antibody. **D–F** 'Wound healing assay' done on near-confluent cultures of NBT-II cells. **A**, **D** Cells cultured under standard conditions. They display an intense desmoplakin immunoreactivity, correlating with the presence of desmosomes (**A**). In the 'wound healing assay', uninduced cells are not able to invade the gap produced in the cell culture (**D**). **B**, **E** NBT-II cells exposed to 30 ng/ml FGF-1 are dissociated, as shown by the loss of desmoplakin immunoreactivity from cellular cortex (**B**) and by the gain of motility, assessed by the 'wound healing' assay. **C**, **F** NBT-II cells exposed to 30 ng/ml FGF-1 and 1 μg/ml cholera toxin do not undergo EMT, as inferred from the presence of desmoplakin immunoreactivity at cell borders (**C**) and the lack of cell motility (**F**).

which cells are placed are therefore capable of directing the cells toward the scattering or the mitogenic response. Since cell compaction and the epithelial state of differentiation are maximized in high-density cultures, it is reasonable to speculate that this culture condition can prevent cells from undergoing EMT. By using sparse and confluent cultures as a tool for discriminating the two effects of FGF-1, we have determined that the scattering response is regulated by a signaling pathway different from that involved in mitogenesis [Boyer and Thiery, 1993]. The FGF-1-induced EMT is inhibited by a rise in intracellular cAMP concentrations (fig. 5), which does not pertub FGF-1-mediated entry into cell division. In contrast, both responses are blocked by inhibitors of tyrosine kinases. The inhibitory role of cAMP has also been demonstrated in EGF-induced cell scattering of rat liver epithelial cells [Manske et al., 1990].

The signaling pathways generated by various scattering factors might vary according to the type of inducing molecule. In that respect, the EMT occurring during embryonic heart formation [Markwald et al., this vol.] is controlled by the activity of a member of the TGF-β family [Potts et al., 1991]. This process is positively regulated by tyrosine kinases but, in contrast to the FGF-1-induced EMT of NBT-II cells, it is not affected by an increase in cAMP levels [Runyan et al., 1990]. Tyrosine kinase activities are generally implicated in cytoskeletal modifications leading to cell motility. This might result from the involvement of $pp125^{FAK}$, a focal contact-associated tyrosine kinase, in signal transduction during cell interactions with ECM [Geiger et al., 1995]. FAK is considered as a point of convergence between integrin and growth factor signaling pathways [Zachary and Rozengurt, 1992]: it is activated by integrin ligation to matrix macromolecules and by a number of growth factors [Zachary et al., 1992; Hordijk et al., 1994; Zachary and Rozengurt, 1992; Seufferlein and Rozengurt, 1994]. Whether or not activation of FAK by each of these stimuli serves a distinct function remains unclear. Interestingly, integrin-mediated activation of FAK induces its association with GRB2 adaptor protein and with c-src protein kinase, resulting in the activation of mitogen-activated protein (MAP) kinase [Schlaepfer et al., 1994]. This might be the point at which integrin- and mitogen-mediated pathways converge. The association of FAK with the Src family of tyrosine kinases can induce the targeting of Src molecules to focal contacts, where they are preferentially localized [Cobb et al., 1994].

It is noteworthy that HGF/SF stimulates the Ras-guanine nucleotide exchanger [Graziani et al., 1993], which is required for the scattering activity of SF [Hartmann et al., 1994]. In agreement with these results, Ras activity is continuously required throughout the process of endothelial cell motility [Fox et al., 1994]. Taken together, these data suggest that Ras activity, stimulated by either growth factor binding to the cell surface or integrin engagement might be a central regulator of EMT processes.

Phospholipase C, another signaling molecule that binds to a number of activated growth factor receptors appears to be required in EGF-induced motility of NR6 fibroblastic cells [Chen et al., 1994], but not in FGF-mediated motility of L6 myoblasts or CHO fibroblasts [Clyman et al., 1994].

A signaling pathway thus appears to be required for EMT to proceed. The specificity of it, as opposed to signals leading to mitogenesis, remains obscure. It is noteworthy that the same type of cellular decision applies in the models in which mitogenesis and cell differentiation can be alternatively induced by the same growth factors. During the stimulation of PC12 cells by insulin or EGF, MAP kinase is activated in the two pathways leading to either cell division or neuronal differentiation. However, the intensity and duration of MAP kinase activation, associated with the entry of the enzyme into the nucleus, is high in insulin or EGF-mediated cell differentiation and weak during mitogenesis. The duration of MAP kinase activation is directly controlled by the intensity of the activation of insulin or EGF receptors, and can be modulated by modifiying the number of receptors expressed at the cell surface [Dikic et al., 1994; Traverse et al., 1994]. A direct modulation of the signals generated at the plasma membrane level by the receptors themselves may also contribute to the dual response to scatter/growth factors.

Is EMT Regulated by Master Genes?

EMT clearly requires transcriptional activity in order to be achieved. Gene expression is modified to a large extent. Figure 6 shows that FGF-1 stimulation of NBT-II cells leads to profound changes in protein expression, as inferred from 2-D gel analysis of the patterns of proteins extracted from control and stimulated cells (part A), and from Western blot analysis of cytoskeletal proteins (part B) and of integrins (part C). In addition, several secreted proteins are specifically associated with the EGF-induced fibroblastic state of differentiation [Bade and Feindler, 1988]. For example, the secretion of collagenases is rapidly induced during EMT [Gavrilovic et al., 1990]. Given the co-ordination and large extent of changes in gene expression, it is tempting to assume that a group of genes under control of a master gene exists for the mesenchymal phenotype, and/or another for the epithelial phenotype [Hay, 1995]. EMT would therefore be associated with switching on the mesenchymal program and/or switching off the epithelial program.

The nature of these master genes, if they exist, remains elusive. E-cadherin, which is capable by itself of inducing some aspects of epithelial differentiation [Mége et al., 1988; McNeill et al., 1990] does not bring about the full range of transformations of mesenchyme to epithelium and therefore cannot be considered as the single master gene controlling the epithelial program. A recent report has shown that the function of *Pax-2,* a transcription factor belonging to the paired-box *(Pax)*-containing family of morphogenetic regulatory genes is required for the MET arising in the developing kidney [Rothenpieler and Dressler, 1993; Davies, this vol.]. *Pax-2* might be important for the condensation of kidney mesenchyme and, like other *Pax* genes, for the subsequent delineation of the boundaries between induced and noninduced cells. However, *Pax-2* is not the only gene controlling MET, since *N-myc* is also an important factor for early stages of epithelium formation [Stanton et al., 1992]. The same kind of genetic cascade, controlling a complex series of events might also operate in EMT. In that respect, *Slug,* a zinc finger-transcription factor homologous to the *Drosophila snail* gene, is expressed in epiblast cells, shortly before they emerge and migrate as mesenchyme during the process of gastrulation. *Slug* is also expressed by precursors of neural crest cells shortly before the onset of cell emigration from the neural tube. Moreover, treatment of embryos with *Slug* antisense oligonucleotides results in the impairment or lack of EMT of NCC and of mesodermal cells [Nieto et al., 1994]. Thus, *Slug* appears as a key factor in the regulation of EMT, at least in these situations. In addition, the activation of the product of the cellular immediate-early gene c-*Fos,* but not of c-*Myc,* is capable of triggering the EMT of normal mammary epithelial cells [Reichmann et al., 1992]. Therefore, it is tempting to postulate that EMT is not under control of a single master gene, and that a series of genetic modifications, temporally co-ordinated, bring about the morphogenetic process of EMT.

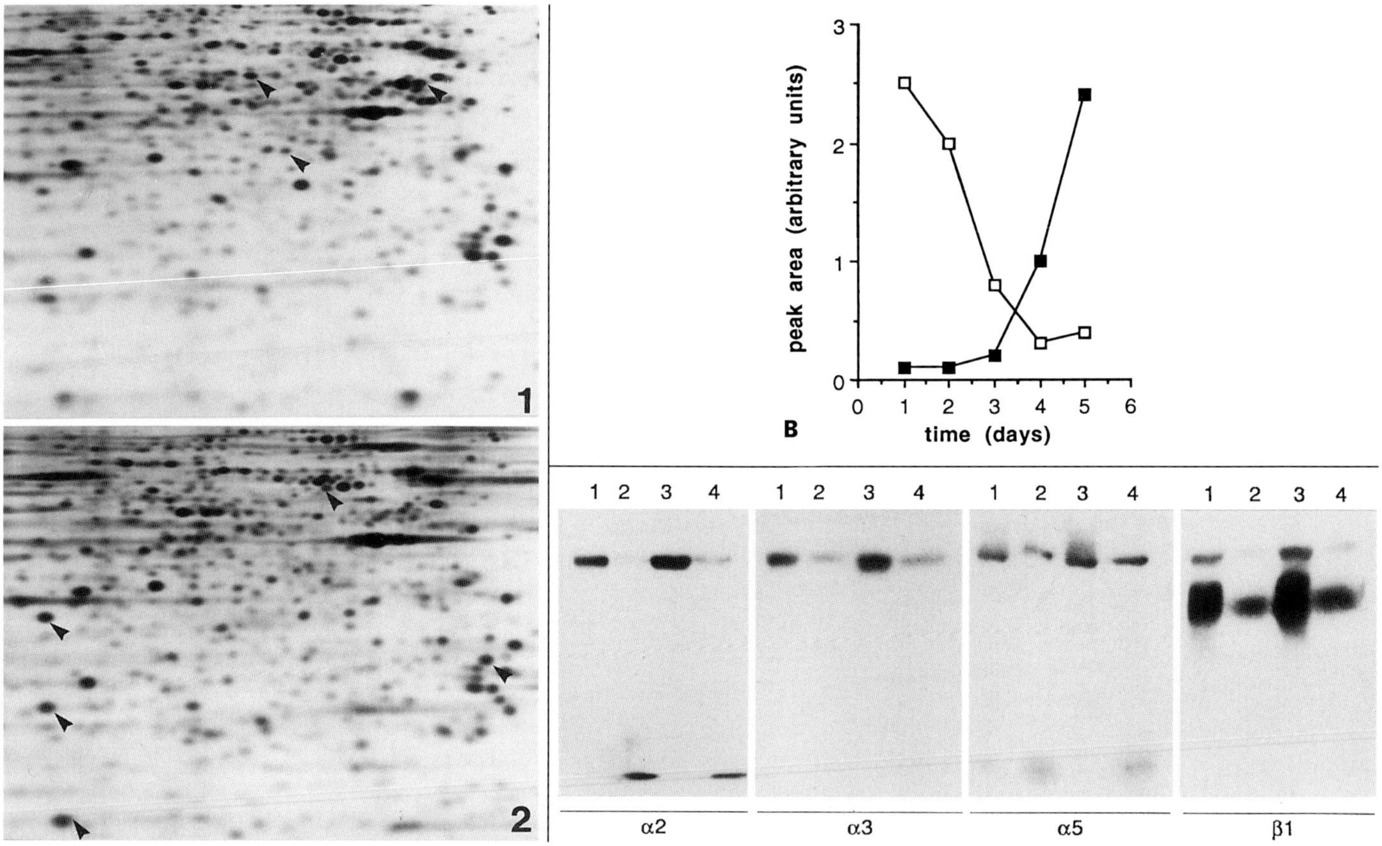

A
C

Fig. 6. FGF-1-induced EMT of NBT-II cells is accompanied by changes in gene expression. **A** Comparison of 2-D gel patterns of [^{35}S-methionine]-labeled proteins extracted from unstimulated cells (upper panel) and cells exposed to 30 ng/ml FGF-1 for 12 h (lower panel). Arrowheads point to spots that are substantially increased under one of the two conditions. Numerous differences between the two gels can be observed, indicating that protein synthesis is modified during EMT. **B** Quantification of cytokeratin and vimentin expression during FGF-1-induced EMT of NBT-II cells. NBT-II cells have been exposed to 30 ng/ml FGF-1 for various periods of time, prior to protein extraction and quantitative immunoblot analysis with cytokeratin antibodies (open squares) and vimentin antibodies (black squares). The steady-state levels of cytokeratin proteins decrease as a function of time, and conversely, vimentin appears 3 days after addition of FGF-1 into the culture medium. **C** Immunoblot analysis of the integrins expressed during the EGF-induced dispersion of normal mammary cells. Lanes 1 and 2 represent proteins extracted in a Triton X-100-soluble fraction (lane 1) and a Triton-insoluble fraction (lane 2) of unstimulated cells. Lanes 3 and 4 are proteins extracted in a Triton-soluble fraction (lane 3) and Triton-insoluble fraction (lane 4) of cells stimulated with EGF for 1 day. Extracted proteins are immunoblotted with antibodies against α2, α3, α5 and β1 integrin subunits. Note that the integrin subunits are upregulated during the EGF-induced EMT.

Concluding Remarks

Studies of in vitro EMTs have led to the characterization of several types of inducing molecules, and to a better understanding of the cellular mechanisms that are involved in the regulation of EMT. However, at present, it is difficult to evaluate the relevance of in vitro EMT-like events to in vivo situations. For example, evidence that inducers of in vitro EMTs can operate in the animal are still missing.

The situation is even more complicated by the fact that in vivo EMTs are multiple; they occur at various times and locations, and affect cells of different origins. Hence, their regulation is likely to depend on the context in which they arise. In vitro examination of EMT events cannot take these subtle differences into account. Therefore, it is of fundamental importance to test the role and the hierarchy of the mechanisms elucidated in in vitro studies using appropriate mutant or transgenic animals which can now be generated.

Acknowledgements

We are indebted to D. Morineau for his excellent photographic work.

The work in the authors' laboratory is supported by the Centre National de la Recherche Scientifique (CNRS), the Association pour la Recherche contre le Cancer (ARC 64-65), the Ligue Française contre le Cancer (National Committee and Committee of Paris), the National Cancer Institute of the NIH (2R01 CA 49417-04), and the Human Frontier Science Program Organization.

References

Bade, E.G., S. Feindler (1988) Liver epithelial cell migration induced by epidermal growth factor or transforming growth factor alpha is associated with changes in the gene expression of secreted proteins. In vitro Cell Dev Biol *24:* 149–154.

Bade, E.G., B. Nitzgen (1985) Extracellular matrix (ECM) modulates the EGF-induced migration of liver epithelial cells in serum-free, hormone-supplemented medium. In vitro Cell Dev Biol *21:* 245–248.

Barrandon, Y., H. Green (1987) Cell migration is essential for sustained growth of keratinocyte colonies: The roles of transforming growth factor-alpha and epidermal growth factor. Cell *50:* 1131–1137.

Behrens, J., L. Vakaet, R. Friis, E. Winterhager, F. Van Roy, M.M. Mareel, W. Birchmeier (1993) Loss of epithelial differentiation and gain of invasiveness correlate with tyrosine phosphorylation of the E-cadherin/β-catenin complex in cells transformed with a temperature-sensitive v-SRC gene. J Cell Biol *120:* 757–766.

Benali, R., F. Dupuit, M. Chevillard, J. Jacquot, B. Haye, E. Puchelle (1991) Modulations of the epithelial phenotype and functional activity of cultured bovine tracheal gland cells: Dependence on the culture medium and passage number. Biol Cell *73:* 49–56.

Berdichevsky, F., C. Gilbert, M. Shearer, J. Taylor-Papdimitriou (1992) Collagen-induced rapid morphogenesis of human mammary epithelial cells: The role of the α2β1 integrin. J Cell Sci *102:* 437–446.

Blay, J., K.D. Brown (1985) Epidermal growth factor promotes the chemotactic migration of cultured rat intestinal epithelial cells. J Cell Physiol *124:* 107–112.

Boyer, B., S. Dufour, J.P. Thiery (1991) E-cadherin expression during the acidic FGF-induced dispersion of a rat bladder carcinoma cell line. Exp Cell Res *201:* 347–357.

Boyer, B., J.P. Thiery (1993) Cyclic AMP distinguishes between two functions of acidic FGF in a rat bladder carcinoma cell line. J Cell Biol *120:* 767–776.

Boyer, B., G.C. Tucker, A.M. Vallés, W.W. Franke, J.P. Thiery (1989) Rearrangements of desmosomal and cytoskeletal proteins during the transition from epithelial to fibroblastoid organization in cultured rat bladder carcinoma cells. J Cell Biol *109:* 1495–1509.

Bradley, R.S., P. Cowin, A.M.C. Brown (1993) Expression of Wnt-1 in PC12 cells results in modulation of plakoglobin and E-cadherin and increased cellular adhesion. J Cell Biol *123:* 1857–1865.

Bringuier, P.P., R. Umbas, H.E. Schaafsma, H.F.M. Karthaus, F.M.J. Debryune, J.A. Schalken (1993) Decreased E-cadherin immunoreactivity correlates with poor survival in patients with bladder tumors. Cancer Res *53:* 3241–3245.

Burridge, K., K. Fath, T. Kelly, G. Nuckolls, C. Turner (1988) Focal adhesions: Transmembrane junctions between the extracellular matrix and the cytoskeleton. Ann Rev Cell Biol *4:* 487–525.

Chan, B.M.C., N. Matsuura, Y. Takada, B.R. Zetter, M.E. Hemler (1991) In vitro and in vivo consequences of VLA-2 expression on rhabdomyosarcoma cells. Science *251:* 1600–1602.

Chen, P., H. Xie, M.C. Sekar, K. Gupta, A. Wells (1994) Epidermal growth factor receptor-mediated cell motility: Phospholipase C activity is required, but mitogen-activated protein kinase activity is not sufficient for induced cell movement. J Cell Biol *127:* 847–857.

Clyman, R.I., K.G. Peters, Y.Q. Chen, J. Escobedo, L.T. Williams, H.E. Ives, E. Wilson (1994) Phospholipase Cγ activation, phosphotidylinositol hydrolysis, and calcium mobilization are not required for FGF receptor-mediated chemotaxis. Cell Adhes Commun *1:* 333–342.

Cobb, B.S., M.D. Schaller, T.-H. Leu, J.T. Parsons (1994) Stable association of $pp60^{src}$ and $pp60^{fyn}$ with the focal adhesion-associated protein tyrosine kinase, $pp125^{FAK}$. Mol Cell Biol *14:* 147–155.

Delannet, M., J.L. Duband (1992) Transforming growth factor-β control of cell-substratum adhesion during avian neural crest cell emigration in vitro. Development *116:* 275–287.

Diamond, M.S., T.A. Springer (1994) The dynamic regulation of integrin adhesiveness. Curr Opin Cell Biol *4:* 506–517.

Dikic, I., J. Schlessinger, I. Lax (1994) PC12 cells overexpressing the insulin receptor undergo insulin-dependent neuronal differentiation. Curr Opin Cell Biol *4:* 702–708.

Duband, J.L., F. Monier, M. Delannet, D. Newgreen (1995) Epithelium-mesenchyme transition during neural crest development. Acta Anat *154:* 63–78.

Dulbecco, R., M. Henahan, M. Bowman, S. Okada, H. Battifora, M. Unger (1981) Generation of fibroblast-like cells from cloned epithelial mammary cells in vitro: A possible new type. Proc Nat Acad Sci USA *78:* 2345–2349.

El Badry, O.M., C. Minniti, E.C. Kohn, P.J. Houghton, W.H. Daughaday, L.J. Helman (1990) Insulin-like growth factor II acts as an autocrine motility factor in human rhabdomyosarcoma tumors. Cell Growth Different *1:* 325–331.

Erickson, C.A. (1990) Cell migration in the embryo and adult organism. Curr Opin Cell Biol *2:* 67–74.

Fabre, M., A.G. de Herreros (1993) Phorbol ester-induced scattering of HT-29 human intestinal cancer cells is associated with down-modulation of E-cadherin. J Cell Sci *106:* 513–522.

Fox, P.L., G. Sa, S.F. Dobrowolski, D.W. Stacey (1994) The regulation of endothelial cell motility by p21*ras*. Oncogene *9:* 3519–3526.

Gamallo, C., J. Palacios, A. Suarez, A. Pizarro, P. Navarro, M. Quintanilla, A. Cano (1993) Correlation of E-cadherin expression with differentiation grade and histological type in breast carcinoma. Am J Pathol *142:* 987–993.

Gavrilovic, J., G. Moens, J.P. Thiery, J. Jouanneau (1990) Expression of transfected transforming growth factor alpha induces a motile fibroblast-like phenotype with extracellular matrix-degrading potential in a rat bladder carcinoma cell line. Cell Regul *1:* 1003–1014.

Geiger, B., S. Yehuda-Levenberg, A.D. Bershadsky (1995) Molecular interactions in the submembrane plaque of cell-cell and cell-matrix adhesions. Acta Anat *154:* 46–62.

Geimer, P., E.G. Bade (1991) The epidermal growth factor-induced migration of rat liver epithelial cells associated with a transient inhibition of DNA synthesis. J Cell Sci *100:* 349–355.

Graziani, A., P. dalla Zonca, P.M. Comoglio (1993) Hepatocyte growth factor/scatter factor stimulates the Ras-guanine nucleotide exchanger. J Biol Chem *268:* 9165–9168.

Grotendorst, G.R. (1984) Alteration of the chemotactic response of NIH/3T3 cells to PDGF by growth factors, transformation, and tumor promotors. Cell *36:* 279.

Gumbiner, B.M. (1993) Proteins associated with the cytoplasmic surface of adhesion molecules. Neuron *11:* 551–564.

Hamaguchi, M., N. Matsuyoshi, Y. Ohnishi, B. Gotoh, M. Takeichi, Y. Nagai (1993) P60$^{v\text{-}src}$ causes tyrosine phosphorylation and inactivation of the N-cadherin-catenin cell adhesion system. EMBO J *12:* 307–314.
Hay, E.D. (1993) Extracellular matrix alters epithelial differentiation. Curr Opin Cell Biol *5:* 1029–1035.
Hay, E.D. (1995) An overview of epithelio-mesenchymal transformation. Acta Anat *154:* 8–20.
Hinck, L., W.J. Nelson, J. Papkoff (1994) Wnt-1 modulates cell-cell adhesion in mammalian cells by stabilizing β-catenin binding to the cell adhesion protein cadherin. J Cell Biol *124:* 729–741.
Hordijk, P.L., I. Verlaan, E.J. van Corven, W.H. Moolenaar (1994) Protein tyrosine phosphorylation induced by lysophosphatidic acid in Rat-1 fibroblasts. J Biol Chem *269:* 645–651.
Hugues, F.J., A.G. McCulloch (1991) Quantification of chemotactic response of quiescent and proliferating fibroblasts in Boyden chambers by computer-assisted image analysis. J Histochem Cytochem *39:* 243–246.
Hynes, R.O. (1992) Integrins: Versatility, modulation, and signaling in cell adhesion. Cell *69:* 11–25.
Juliano, R.L., J.A. Varner (1993) Adhesion molecules in cancer: The role of integrins. Curr Opin Cell Biol *5:* 812–818.
Kawanishi, J., J. Kato, K. Sasaki, S. Fujii, N. Watanabe, Y. Niitsu (1995) Loss of E-cadherin-dependent cell-cell adhesion due to mutation of the β-catenin gene in a human cancer cell line, HCS-39. Mol Cell Biol *15:* 1175–1181.
Klein, C.E., T. Steinmayer, D. Kaufmann, L. Weber, E.-B. Bröcker (1991) Identification of a melanoma progression antigen as integrin VLA-2. J Invest Dermatol *96:* 281–284.
Klemke, R.L., M. Yebra, E.M. Bayna, D.A. Cheresh (1994) Receptor tyrosine kinase signaling required for integrin αVβ5 cell motility but not adhesion on vitronectin. J Cell Biol *127:* 859–866.
Lin, C.Q., M.J. Bissell (1993) Multi-faceted regulation of cell differentiation by extracellular matrix. FASEB J *7:* 737–743.
Manske, M., S. Feindler, E.G. Bade (1990) The epidermal growth factor-induced cell migration and expression of the 47,000 M_r secreted glycoprotein EIP-1 of rat liver epithelial cells are down-regulated by cyclic AMP. Eur J Cell Biol *52:* 201–206.
Matsuyoshi, N., M. Hamaguchi, S. Taniguchi, A. Nagafuchi, S. Tsukita, M. Takeichi (1992) Cadherin-mediated cell-cell adhesion is perturbed by *v-src* tyrosine phosphorylation in metastatic fibroblasts. J Cell Biol *118:* 703–714.
Matthay, M.M., J.P. Thiery, F. Lafont, M.F. Stampfer, B. Boyer (1993) Transient effect of epidermal growth factor on the motility of an immortalized mammary epithelial cell line. J Cell Sci *106:* 869–878.
Mayer, B., J.P. Johnson, F. Leitl, K.W. Jauch, M.M. Heiss, F.W. Schildberg, W. Birchmeier, I. Funke (1993) E-cadherin expression in primary and metastatic gastric cancer: Down-regulation correlates with cellular dedifferentiation and glandular disintegration. Cancer Res *53:* 1690–1695.
McNeill, H., M. Ozawa, R. Kemler, W.J. Nelson (1990) Novel function of the cell adhesion molecule uvomorulin as an inducer of cell surface polarity. Cell *62:* 309–316.
Mége, R.M., F. Matsuzaki, W. Gallin, J.I. Goldberg, B. Cunningham, G.M. Edelman (1988) Construction of epithelioid sheets by transfection of mouse sarcoma cells with cDNAs for chicken cell adhesion molecules. Proc Nat Acad Sci USA *85:* 7274–7278.
Miettinen, P.J., R. Ebner, A.R. Lopez, R. Derynck (1994) TGF-β induced transdifferentiation of mammary epithelial cells to mesenchymal cells: Involvement of type I receptors. J Cell Biol *127:* 2021–2036.
Neugebauer, K.M., K.J. Tomaselli, J. Lilien, L.F. Reichardt (1988) N-cadherin, NCAM, and integrins promote retinal neurite outgrowth on astrocytes in vitro. J Cell Biol *107:* 1177–1187.
Nieto, M.A., M.G. Sargent, D.G. Wilkinson, J. Cooke (1994) Control of cell behavior during vertebrate development by Slug, a zinc finger gene. Science *264:* 835–839.
Oka, H., H. Shiozaki, K. Kobayashi, M. Inoue, H. Tahara, T. Kobayashi, Y. Takatsuka, N. Matsuyoshi, S. Hirano, M. Takeichi, T. Mori (1993) Expression of E-cadherin cell adhesion molecules in human breast cancer tissues and its relationship to metastasis. Cancer Res *53:* 1696–1701.
Potts, J.D., J.M. Dagle, J.A. Walder, D.L. Weeks, R.B. Runyan (1991) Epithelial-mesenchymal transformation of embryonic cardiac endothelial cells is inhibited by a modified antisense oligodeoxynucleotide to transforming growth factor β3. Proc Nat Acad Sci USA *88:* 1516–1520.
Potts, J.D., R.B. Runyan (1989) Epithelial-mesenchymal cell transformation in the embryonic heart can be mediated, in part, by transforming growth factor-β. Dev Biol *134:* 392–401.
Prieto, A.L., K.L. Crossin (1995) Cell-cell adhesion molecules in epithelial-mesenchymal transformations. Acta Anat *154:* 21–33.
Reynolds, A.B., J. Daniel, P.D. McCrea, M.J. Wheelock, J. Wu, Z. Zhang (1994) Identification of a new catenin: the tyrosine kinase substrate p120^{cas} associates with E-cadherin complexes. Mol Cell Biol *14:* 8333–8342.
Reynolds, A.B., L. Herbert, J.L. Cleveland, S.T. Berg, J.R. Gaut (1992) p120, a novel substrate of protein tyrosine kinase receptors and of p60$^{v\text{-}src}$, is related to cadherin-binding factors β-catenin, plakoglobin and *armadillo.* Oncogene *7:* 2439–2445.
Rosen, E.M., S.K. Nigam, I.D. Goldberg (1994) Scatter factor and the c-met receptor: A paradigm for mesenchymal/epithelial interaction. J Cell Biol *127:* 1783–1787.
Rothenpieler, U.W., G.R. Dressler (1993) *Pax-2* is required for mesenchyme-to-epithelium conversion during kidney development. Development *119:* 711–720.
Rubinfeld, B., B. Souza, I. Albert, O. Müller, S.H. Chamberlain, F.R. Masiarz, S. Munemitsu, P. Polakis (1993) Association of the APC gene product with β-catenin. Science *262:* 1731–1734.
Runyan, R.B., J.D. Potts, R.V. Sharma, C.P. Loeber, J.J. Chiang, R.C. Bhalla (1990) Signal transduction of a tissue interaction during embryonic heart development. Cell Regul *1:* 301–313.
Schiro, J.A., B.M.C. Chan, W.T. Roswit, P.D. Kassner, A.P. Pentland, M.E. Hemler, A.Z. Eisen, T.S. Kupper (1991) Integrin α2β1 (VLA-2) mediates reorganization and contraction of collagen matrices by human cells. Cell *67:* 403–410.
Schlaepfer, D.D., S.K. Hanks, T. Hunter, P. van der Geer (1994) Integrin-mediated signal transduction linked to Ras pathway by GRB2 binding to focal adhesion kinase. Nature *372:* 786–791.
Seufferlein, T., E. Rozengurt (1994) Lysophosphatidic acid stimulates tyrosine phosphorylation of focal adhesion kinase, paxillin, and p130. J Biol Chem *269:* 9345–9351.
Shackelford, G.M., K. Willert, J. Wang, H.E. Varmus (1993) The Wnt-1 proto-oncogene induces changes in morphology, gene expression, and growth factor responsiveness in PC12 cells. Neuron *11:* 865–875.
Shibamoto, S., M. Hahakawa, K. Takeuchi, T. Hori, N. Oku, K. Miyazawa, N. Kitamura, M. Takeichi, F. Ito (1994) Tyrosine phosphorylation of β-catenin and plakoglobin enhanced by hepatocyte growth factor and epidermal growth factor in human carcinoma cells. Cell Adhes Commun *1:* 295–305.
Shimonaka, M., Y. Yamaguchi (1994) Purification and biological characterization of epitaxin, a fibroblast-derived motility factor for epithelial cells. J Biol Chem *269:* 14284–14289.
Shimoyama, Y., A. Nagafuchi, S. Fujita, M. Gotoh, M. Takeichi, S. Tsukita, S. Hirohashi (1992) Cadherin dysfunction in a human cancer cell line: possible involvement of loss of α-catenin expression in reduced cell-cell adhesiveness. Cancer Res *52:* 5770–5774.
Shiota, G., D.B. Rhoads, T.C. Wang, T. Nakamura, E.V. Schmidt (1992) Hepatocyte growth factor inhibits growth of hepatocellular carcinoma cells. Proc Nat Acad Sci USA *89:* 373–377.
Sommers, C., E.P. Gelmann, R. Kemler, P. Cowin, S.W. Byers (1994) Alterations in β-catenin phosphorylation and plakoglobin expression in human breast cancer cells. Cancer Res *54:* 3544–3552.
Stampfer, M.R., J.C. Bartley (1988) In Lippmann M, R Dickson (eds): Breast Cancer: Cellular and Molecular Biology. Boston, Kluwer Academic Publishers, pp 1–24.
Stanton, B.R., A.S. Perkins, L. Tessarollo, D.A. Sassoon, L.F. Parada (1992) Loss of *N-myc* function results in embryonic lethality and failure of the epithelial component of the embryo to develop. Genes Dev *6:* 2235–2247.

Stracke, M.L., E.C. Kohn, S.A. Aznavoorian, L.L. Wilson, D. Salomon, H.C. Krutzsch, L.A. Liotta, E. Schiffman (1988) Insulin-like growth factors stimulate chemotaxis in human melanoma cells. Biochem Biophys Res Commun *153:* 1076–1083.

Su, L.-K., B. Vogelstein, K.W. Kinzler (1993) Association of the APC tumor suppressor protein with catenins. Science *262:* 1734–1737.

Traverse, S., K. Seedorf, H. Paterson, C.J. Marshall, P. Cohen, A. Ullrich (1994) EGF triggers neuronal differentiation of PC12 cells that overexpress the EGF receptor. Curr Opin Cell Biol *4:* 694–701.

Tsarfaty, I., S. Rong, J.H. Resau, S. Rulong, P. Pinto da Silva, G.F. Vande Woude (1994) The met proto-oncogene mesenchymal to epithelial cell conversion. Science *263:* 98–101.

Tucker, G.C., B. Boyer, J. Gavrilovic, H. Emonard, J.P. Thiery (1990) Collagen-mediated dispersion of NBT-II rat bladder carcinoma cells. Cancer Res *50:* 129–137.

Tucker, G.C., B. Boyer, A.M. Vallés, J.P. Thiery (1991) Combined effects of extracellular matrix and growth factors on NBT-II rat bladder carcinoma cell dispersion. J Cell Sci *100:* 371–380.

Vallés, A.M., B. Boyer, J. Badet, G.C. Tucker, D. Barritault, J.P. Thiery (1990a) Acidic fibroblast growth factor is a modulator of epithelial plasticity in a rat bladder carcinoma cell line. Proc Nat Acad Sci USA *87:* 1124–1128.

Vallés, A.M., G.C. Tucker, J.P. Thiery, B. Boyer (1990b) Alternative patterns of mitogenesis and cell scattering induced by acidic FGF as a function of cell density in a rat bladder carcinoma cell line. Cell Regul *1:* 975–988.

Woolf, A.S., J. Kolatsi-Joannou, P. Hardman, E. Andermacher, C. Moorby, L.G. Fine, P.S. Jat, M.D. Noble, E. Gherardi (1995) Roles of hepatocyte growth factor/scatter factor and the met receptor in the early development of the metanephros. J Cell Biol *128:* 171–184.

Zachary, I., J. Sinnett-Smith, E. Rozengurt (1992) Bombesin, vasopressin, and endothelin stimulation of tyrosine phosphorylation in Swiss 3T3 cells. J Biol Chem *267:* 19031–19034.

Zachary, I., E. Rozengurt (1992) Focal adhesion kinase (p125FAK): A point of convergence in the action of neuropeptides, integrins, and oncogenes. Cell *71:* 891–894.

Author Index Vol. 156, No. 3, 1996

Subject Index Vol. 156, No. 3, 1996